Biology of Fresh Waters

TERTIARY LEVEL BIOLOGY

A series covering selected areas of biology at advanced undergraduate level. While designed specifically for course options at this level within Universities and Polytechnics, the series will be of great value to specialists and research workers in other fields who require knowledge of the essentials of a subject.

Recent titles in the series:

The Lichen-Forming Fungi	Hawksworth and Hill
Social Behaviour in Mammals	Poole
Physiological Strategies in Avian Biology	Philips, Butler and Sharp
An Introduction to Coastal Ecology	Boaden and Seed
Microbial Energetics	Dawes
Molecule, Nerve and Embryo	Ribchester
Nitrogen Fixation in Plants	Dixon and Wheeler
Genetics of Microbes (2nd edn.)	Bainbridge
Seabird Ecology	Furness and Monaghan
The Biochemistry of Energy Utilization in Plants	Dennis
The Behavioural Ecology of Ants	Sudd and Franks
Anaerobic Bacteria	Holland, Knapp and Shoesmith
An Introduction to Marine Science (2nd edn.)	Meadows and Campbell
Seed Dormancy and Germination	Bradbeer
Plant Growth Regulators	Roberts and Hooley
Plant Molecular Biology (2nd edn.)	Grierson and Covey
Polar Ecology	Stonehouse
The Estuarine Ecosystem (2nd edn.)	McLusky
Soil Biology	Wood
Photosynthesis	Gregory
The Cytoskeleton and Cell Motility	Preston, King and Hyams
Waterfowl Ecology	Owen and Black

TERTIARY LEVEL BIOLOGY

Biology of
Fresh Waters

Second Edition

Peter S. Maitland, B.Sc., Ph.D., FRSE

Fish Conservation Centre
Stirling

Blackie
Glasgow and London

Published in the USA by
Chapman and Hall
New York

Blackie and Son Limited
Bishopbriggs, Glasgow G64 2NZ
and
7 Leicester Place, London WC2H 7BP

Published in the USA by
Chapman and Hall
a division of Routledge, Chapman and Hall, Inc.
29 West 35th Street, New York, NY 10001–2291

British Library Cataloguing in Publication Data

Maitland, Peter S. (Peter Salisbury) *1937–*
Biology of fresh waters.
2nd ed
1. Freshwater organisms
I. Title   II. Series
574.929

ISBN 978-94-011-7854-9   ISBN 978-94-011-7852-5 (eBook)
DOI 10.1007/978-94-011-7852-5

Library of Congress Cataloging-in-Publication Data

Maitland, Peter S.
Biology of fresh waters / Peter S. Maitland. — 2nd ed.
p.   cm. — (Tertiary level biology)
Includes bibliographical references and index.
ISBN 978-94-011-7854-9
1. Freshwater biology.   I. Title.   II. Series
QH96.M28   1990
574.92.9–do20
90-2037
CIP

Typesetting by Thomson Press (India) Limited, New Delhi

Preface

In the decade since the first edition of this book was published advances have been made in our knowledge of the fresh waters of the world, especially in understanding many of the processes involved in their functioning as systems and in countering the problems created by human activities. New problems too, many of an international nature, have loomed during this period—of which global warming and the acidification of fresh waters in many parts of the world are notable examples. In addition, much has now been published concerning the aquatic flora, fauna and ecology of previously poorly known geographic areas, notably Australasia.

The second edition of this book is a revision which updates the text in the light of recent advances in our knowledge of freshwater biology. Inevitably, in an elementary volume such as this, the treatment of many of the basic principles and processes remains the same. However, several new sections are included covering a range of topics such as acid deposition and the acidification process, bacterial decomposition and aquaculture. The book includes many new references and suggestions for up-to-date reading in particular topics.

The objective of the second edition remains the same as that of the first. It is intended as a basic introduction to the major aspects of freshwater biology at a level suitable for undergraduates. It should also prove useful, as apparently did the first edition, to professional workers in related fields, e.g. water engineers and chemists, aquaculturists and planners.

In revising this book I have been helped by several of the colleagues mentioned in the Preface to the first edition. In particular I would like to thank Mr James D. Hamilton (formerly of Paisley College of Technology), Dr Johanna Laybourn-Parry (University of Lancaster) and Mr David W Mackay (formerly of the Clyde River Purification Board). The first edition of the book was tested extensively during the period I spent as Senior Lecturer in Ecology at the University of St. Andrews (1982–85) and I am grateful to the late and sadly missed Professor David H.N. Spence for his encouragement at that time.

PETER S. MAITLAND

Contents

6 ADAPTATION TO ENVIRONMENT: STRATEGIES FOR SURVIVAL

7 COMMUNITIES AND ENERGY FLOW

CHAPTER ONE
THE AQUATIC ENVIRONMENT

In many parts of the world, the changing weather patterns of the 1980s certainly focussed attention much more on the importance of fresh water. In times of drought, not only were lakes and rivers shallower than normal but reservoirs were lower, crops were poorer, gardens were drier and cars were dirtier as a result. Yet elsewhere, or at another time of year, exceptionally heavy rainfalls have created havoc with flooding and subsequent damage to people and property.

In many countries of the world there is an abundance of fresh water, and with all this water about, there should be enough for every need; yet there is an enormous range of demands—often conflicting ones. In addition, humans do many things to affect the water even before it gets into lakes and rivers—land use in particular is of major importance in this respect. The use of fertilisers, herbicides and pesticides in agriculture and forestry means that many of these chemicals or their residues end up in water courses, the process often being aided by the draining and ditching of land so that water flows off faster.

In many communities, humans use over 100 l of water a day for domestic needs: drinking, cooking, washing, bathing, toileting and so on. Industry too has huge demands. Wastes from domestic sewage and industry are passed into rivers (sometimes with poor treatment) to be carried away to the sea. Hydro-electric schemes have harnessed many large rivers by the construction of dams and tunnels. Finally, humans use water for recreation: picnicking beside it, birdwatching over it, paddling and bathing in it, boating on it and fishing in it.

Naturally, with all these pressures there are problems, both for humans and for wildlife. Many waters have been so misused that they are unfit for either. Over-enrichment from agricultural fertilisers has caused algal blooms on many lowland lakes; when these blooms reach a maximum and die, fish kills result. Domestic and industrial pollution can be even more severe and may completely eliminate all aquatic plants and animals, as has happened in the lower parts of many of the world's rivers. These stretches of extreme pollution, devoid of oxygen or poisoned by chemicals, also act as a barrier to migrating fish which thus cannot occupy stretches upstream, however clean these may be. Simple weirs and dams on a river do this too. Aerial pollution also has had a devastating effect on many waters in the northern hemisphere: where 100 years ago there were healthy salmonid populations, thousands of waters are now too acid for any fish to survive.

The larger waters tend often to hold the limelight, yet one of the saddest things over the last hundred years is what has happened to smaller water bodies—especially in lowland areas. In many parts of the world, virtually all the small streams in lowland agricultural areas have been ditched, canalised and piped to such an extent that there are virtually no natural examples left. With small ponds, the situation is even worse, for there has been a gradual draining and filling-in of such waters over the last century, and even the few which are still left are degraded or threatened.

As a background to such problems, this book attempts to explain the basic principles of fresh water as a medium, and in particular as an environment in which communities of plants and animals thrive. Figure 1.1 illustrates the variety of techniques involved in modern investigative freshwater biology. This initial chapter is devoted to the main physical and chemical features of fresh water as an environment. Factors related to these properties, but more or less peculiar to either standing water (e.g. stratification) or running water (e.g. unidirectional current), are dealt with in detail later. All freshwater bodies are dynamic systems, and not only are their organisms affected by the physicochemical conditions there, but also the plants and animals interact and may influence both the habitat and one another. It is therefore essential to consider the organisms as part of the environment. Initially, only major influences of the organisms on the physical and chemical conditions of the environment are reviewed. Inter- and intraspecific relationships among plants and animals are dealt with later in the chapters pertaining to communities.

The fresh waters of the world are unimportant compared on an area basis with most land and sea surfaces. However, some of the largest rivers and lakes are of impressive dimensions and are of major importance in the general ecology and cycling relationships of the regions in which they occur. Fresh water itself and many freshwater systems are of fundamental importance to humans, and this topic is treated in some detail in a later chapter.

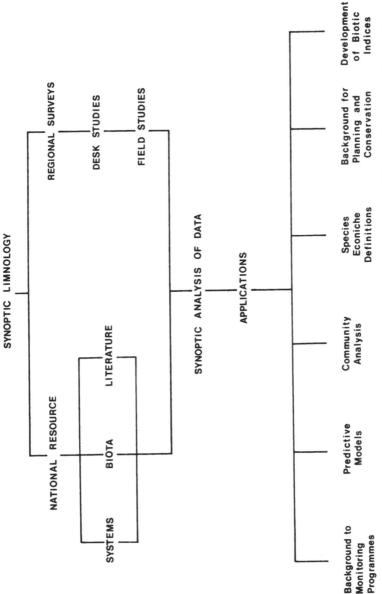

Figure 1.1 A network diagram illustrating a modern approach to investigative freshwater biology with a variety of user endpoints integrating monitoring, planning and conservation (from Maitland, 1979).

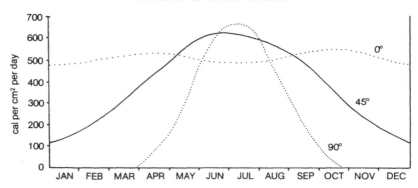

Figure 1.2   The average quantities of energy received from direct solar radiation at different latitudes.

1.1   Physics

1.1.1   *Radiant energy and optics*

The major source of energy in most fresh waters enters as solar radiation. The total amount of radiation reaching the surface of a water depends on time of year, geography, altitude, state of the atmosphere and several other (usually local) factors. Near the equator, with no cloud cover, some 515 cal/cm^2 per day are received, with little variation throughout the year (about $\pm$ 35 cal/cm^2). At the North Pole, however, almost 670 cal/cm^2 are received at midsummer, but none around midwinter (Figure 1.2). The amount of radiation received at any latitude increases with altitude.

Light falling on water undergoes changes which depend on both the angle at which it meets the surface and the nature of the water itself. Rays meeting the surface may be partly reflected and partly transmitted—the latter then becoming more vertical (Figure 1.3). The degree of reflection depends on the angle of incidence, and increases as it moves from the perpendicular. Thus, both diurnally and annually, the amount of energy reflected from the surface of lakes and rivers varies greatly according to the height of the sun. This results in shorter dawns and dusks below water than above; moreover, when the sun is low and the angle of incidence greater, the increasing distance per unit depth through which the rays must pass under water means that they are extinguished even more rapidly. As well as being reflected from the surface, some radiation is scattered

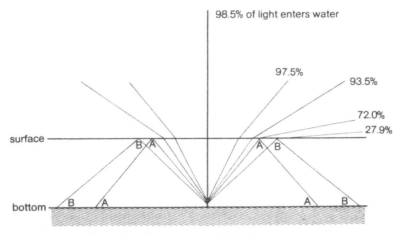

Figure 1.3 The refraction of light through water.

from below the water into the atmosphere. In normal waters the amounts involved in each loss are about the same.

Apart from direct sunlight, the amount of which is dependent on cloud cover, there is also some indirect solar radiation received from the sky. This may account for about 20% of the radiation reaching the water surface, and normally less than 20% of it is reflected.

The radiant energy passing through the surface is further altered: part is absorbed by the water, by its suspended materials and solutes, and transformed to heat, and part is dispersed. Quite apart from the effects of impurities, such as suspended and dissolved materials which may affect transparency (Figure 1.4), there are variations in the transmissions of different wavelengths of light, even through pure water. Short wavelengths travel further than long ones: minimum absorption is at 4700 Å in the blue region, so that in pure water objects at some depth appear blue. Long wavelengths (orange and red rays) are absorbed rapidly: some 90% of wavelengths greater than 7500 Å are absorbed within 1 m in pure water; thus infra-red rays can rarely penetrate far.

Most dissolved solids absorb short wavelengths strongly, and long wavelengths least. When suspended solids are present in small amounts, water is most transmissive to green light; when large amounts of materials occur, transmission is greatest for the longer orange and red wavelengths. As water transparency decreases, the limit of detectable light (whose wavelength gets progressively longer) approaches the surface. In most

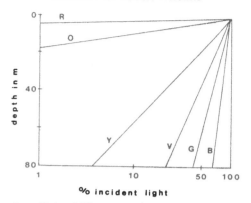

Figure 1.4   Attenuation of light of different wavelengths in pure water (R = red; O = orange; Y = yellow; V = violet; G = green; B = blue).

situations radiation tends to decrease logarithmically with depth.

The observed colour of natural fresh waters may be due to their actual colour, or an apparent colour made up from this and the influence of other factors. The quality of the incident light, the selective transmission of wavelengths, the amount and quality of suspended matter, and (in shallow water) the colour of the substrate, are all important. Basically, the colour observed on looking into a river or lake is that of the upward scattered light. Natural systems where the water is relatively pure appear very dark, unless they are very shallow (where the colour of the substrate is important) or reflect colour from their surroundings. In waters where there are large quantities of suspended materials, either living (e.g. algae) or inanimate (e.g. clay), coloured light is reflected and combines with the transmission effect to give a particular colour. In natural waters where there are small amounts of suspended matter, the water looks green, but where there are large amounts the water looks yellow or brown. Of the dissolved substances which influence the selective transparency, humus materials (usually yellow or brown) are important; thus deep waters which are very pure appear dark-bluish, but those affected by humic materials may be dark-brown or black. In shallow waters of these types the colour of the substrate has a major influence, whereas in waters with large amounts of suspended materials both depth and substrate are relatively unimportant.

In temperate and arctic areas, ice can be a major factor in controlling the amount of radiation entering a water body. The transparency of absolutely pure ice is very high, and with such 'black' ice a relatively high proportion of the solar radiation penetrates below. Where, however, the ice is less transparent (usually where it has been formed under rough conditions) or covered by snow or dirt, very little energy may reach the water underneath due to loss by reflection or absorption.

Though the specific heat of ice is low (0.5), that of water is high (1.0), and indeed much higher than of many other materials. This property, defined as the number of calories of heat needed to raise the temperature of a unit mass of a material by 1°C, compared with that for water, means that temperature conditions in water are much more stable than those in air: rapid diurnal and seasonal changes of temperature in natural waters are rare—an important factor in the ecology of many aquatic organisms.

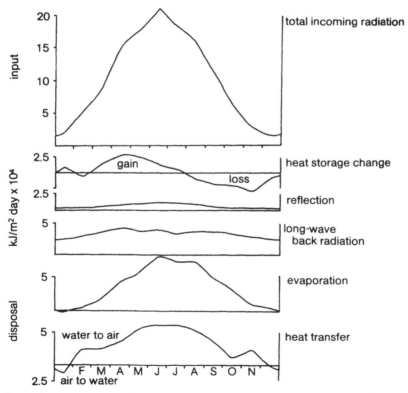

Figure 1.5 Average cycle of radiation balance components in Loch Leven for the period 1968–71 (after Smith, 1974).

In any comprehensive study of a water it is important to construct a heat budget, and to consider how this varies with season (Figure 1.5). Most heat from solar radiation accumulates directly by absorption, though some may be conducted from the air or from the earth beneath. The latter is important for subterranean waters. In some circumstances heat is available from the condensation of water vapour near the surface. The main source of heat loss is through radiation at the surface, though losses through evaporation and conduction (at the surface and to the substrate below) are also significant.

1.1.2   Density and thermal properties

The earth is unusual among planets in having large amounts of water on its surface. Water itself is remarkable in being almost the only major material existing as a liquid on the earth's surface at ordinary pressures. At 0°C and atmospheric pressure (760 mmHg) the density of pure water is more than 700 times that of air. This means that the tissues of aquatic organisms need far less support than those of terrestrial ones, and there can be a great reduction in skeletal structure—this is especially valuable to large animals. The density of water in different places can vary in time and space, and even small variations are important. They are mainly due to temperature and dissolved solids, especially the former.

Unlike most materials, whose density increases with decreasing temperature, water reaches a maximum density (1.000) at about 4°C (actually 3.94°C). This critical temperature (Figure 1.6) varies according to pressure (an increase of 10 atmospheres decreases it by 0.2°C) and salinity (an increase of 1% decreases it by 0.2°C). Below 4°C, water decreases gradually in density to its freezing point, when the density decreases sharply—ice is more than 8% less dense than water at 0°C. This anomalous temperature/density relationship is due to the differential packing of water molecules at different temperatures.

This anomalous change in density is of considerable significance biologically, as will become clear in succeeding chapters. After cooling, water starts freezing at the surface, but near the bottom the temperature may be much greater—usually about 4°C. The actual formation of ice cover depends on a variety of factors. The most critical of these are that the water temperature in the whole water body is 4°C or below, and that the weather is clear with little wind. Conditions for the break-up of ice cover are the opposite of these.

The effects of an increase in temperature on the density of water are

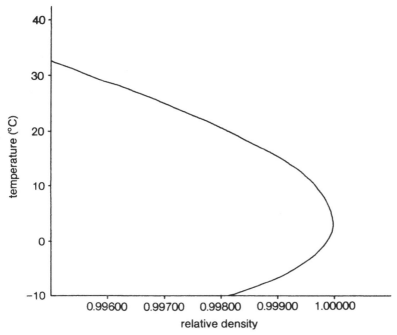

Figure 1.6 The relationship between the relative density and the temperature of pure water.

also of significance biologically. Density changes much more rapidly at higher temperatures than lower; thus a 1°C change in temperature at 24°C decreases the density many times more than the same change at 4°C. This is important for buoyancy, as planktonic animals, for instance, will tend to sink more rapidly at higher temperatures than lower ones. This factor is also relevant to the development of stratification. Also, because of the great difference in density between water and air, aquatic animals must overcome much greater resistance than terrestrial ones when moving and must therefore expend more energy for a given return.

The density of water at constant temperature varies with pressure, and increases slowly in a linear fashion with depth. Because of the dual effect of pressure and temperature, the maximum density of water may occur not at 3.94°C but at a lower temperature in very deep lakes; thus it may be less than 3°C at 1000 m. As a result, lakes where cold water can be mixed to great depths may have deep-water temperatures lower than 4°C during summer stratification.

Though not as important as formerly suggested, temperature itself influences the biology of fresh waters. Most freshwater animals and all plants are poikilothermic, and thus their temperature varies with that of their surroundings. The physiological reactions taking place within them (e.g. photosynthesis, respiration, digestion) are biochemical ones, and the rate at which they take place is very dependent on the ambient temperature. Many freshwater organisms react to temperature with considerable precision, and are capable of detecting differences as small as 0.2°C in their surroundings. The life cycles of many temperate organisms are geared to the seasons: the breeding biology of many insects is closely linked with the day-length and/or temperature of their surroundings. Morgan and Waddell (1961), for instance, found that the number of species of chironomid midge emerging from a Scottish loch was directly proportional to water temperature (Figure 1.7).

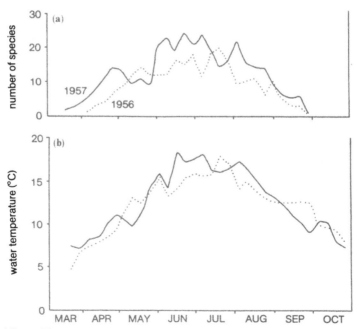

Figure 1.7 (a) The numbers of species of emerging midges (Chironomidae) and (b) water temperatures in Loch Dunmore during 1956 and 1957 (after Morgan and Waddell, 1961).

Figure 1.8 Conditions in fresh waters range from shallow still ponds to enormous turbulent cascading waterfalls such as the one shown here (Photo: P.S. Maitland.)

1.1.3 Movement

Several factors influence the nature and extent of water movements. Because of the importance of winds in producing movements in standing waters, and gradients in causing currents in running waters, local topography is of considerable significance (Figure 1.8). The nature of the basin or channel is also important. In very large bodies of open water currents may be influenced by the rotation of the earth as is the sea.

In open waters the energy producing most currents is derived from the wind, from changes in density within the water, or from kinetic differences in level (e.g. as in a stream channel). The wind can produce currents in two ways: (a) by driving action at the surface and (b) by driving surface water towards one end of the lake (Figure 1.9), thus inducing sub-surface currents in the opposite direction. Convection currents, however, act vertically and are usually caused by sudden cooling and subsequent sinking of surface waters. Currents in channels are related to the gradient of the channel bed, or more precisely to differences in surface level between two points in the channel.

The actual movement of water particles within a current can be of two types: laminar or turbulent. Pure laminar flow is rare in nature, normally occurring at very low velocities. Its main importance in aquatic ecosystems is that of transportation. Most currents generate turbulent motion

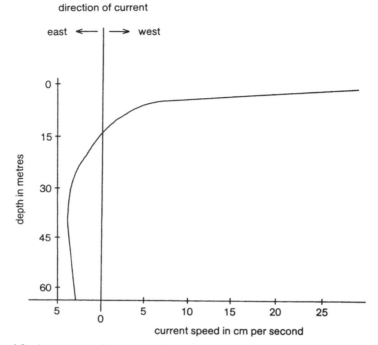

Figure 1.9 A current profile taken in Loch Garry, showing a strong westerly current at the surface (caused by wind action) which decreased in deeper water until about 15 m, where movement changed to a return easterly current (after Wedderburn, 1910).

perpendicular to the main line of current movement. For this reason, since currents are normally horizontal, most turbulence is vertical in nature. Turbulent motion is of great importance in aquatic systems, due to its mixing effect. This is very significant, not only in terms of heat transport and water chemistry, but also in connection with suspended particles—especially living organisms such as bacteria and algae.

The surface motion of water induced by wind has been discussed by Langmuir (1938) and others. The influence of moving waves is restricted mainly to surface waters; the water particles involved rise and fall rhythmically in a path which is almost circular, so that after each cycle every particle returns to very near its original position (Figure 1.10). The major cause of moving waves is the wind, the force of which causes the water particles to move from a state of equilibrium to a point at which they are influenced by gravity, which tends to return them to their original

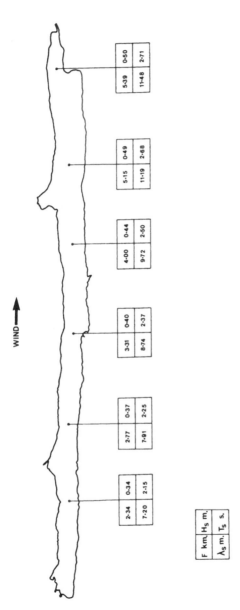

Figure 1.10 Deep water wave characteristics for a large simple lake (Loch Ness). The wind speed is 10 m/s and the arrow indicates the wind direction. The upper left hand figure in each box is the effective fetch, the upper right is the significant wave height, the lower left is the wavelength and the lower right hand is the wave period (from Smith *et al.*, 1981).

position. Surface waves arise when the wind over the water surface reaches a critical force. Most surface waves in deep water travel at a velocity $v = 12.5 \times$ wavelength. If the wavelength is small (less than about 1.6 cm), the velocity increases as the wavelength increases. In small lakes the size of the waves is independent of depth, while in large lakes waves increase with depth. The height of the highest waves on a lake is proportional to the square root of the wind fetch generating them.

The depth to which surface waves can penetrate is of importance biologically. In deep water the amplitude is reduced by 50% for each increase in depth of about 10% of the wavelength. There is no relationship between the length and height of waves, but on average the former is about 20 times the latter. The effect of such waves decreases rapidly below the surface; thus the influence of even very large waves of more than 1 m in height rarely penetrates below 20 m, but up to this depth the spatial distribution of organisms may be affected. In shallow waters where the wavelength is greater than the water depth, the surface motion influences the whole water column, and the extent of the movement is similar from surface to bottom. Such movements are important in the development of substrate types, because regular sedimentation is prevented, and the bottom tends to consist of either clean sand (which is unstable) or stones and rock.

Waves of this type are known as travelling waves. In them the water particles, though they may differ in phase, all move through the same distance. In the type of movement known as a standing wave or seiche (Figure 1.11), the phase of the water particles is the same, but they move through different distances. The forces producing seiches (usually local variations in atmospheric pressure, wind or rainfall) cause differential pressures on the surface of one part of a water body compared with

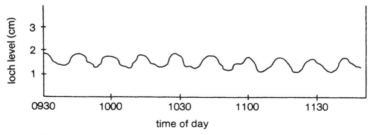

Figure 1.11   Variations in the surface level of Loch Earn due to a seiche (after Chrystal, 1910).

another. Seiches of large amplitudes can be caused by relatively small differences in pressure; moreover, seiches are not confined to surface waters, but can occur at some depth without any movement at the surface. Such standing waves are called internal seiches and occur mainly as underwater oscillations in stratified lakes; they are usually brought about by strong winds, but occasionally by the inflow of a large volume of water to one part of a lake. The water movements caused by seiches are extremely important in lake ecology and are further discussed in Chapter 3.

1.1.4 Suspended solids

The significance of suspended solids in natural waters normally relates to their physical effects there, particularly with regard to their influence on light and bottom sediments (Figure 1.12). In systems affected by humans, however, the effects may be both physical and chemical and suspended solids are regarded as so important in the pollution context that they are routinely measured when monitoring water quality.

Inorganic suspended solids are more important in rivers than in lakes and usually originate from terrestrial sources—especially where there has been soil disturbance of some kind, followed by heavy rainfall (Milner *et al.* 1981). Enormous amounts of materials can be washed into running waters at such times and carried long distances downstream. Where the current slows, the heavier particles settle out to fill in pools but lighter particles may be carried right out to sea and many millions of tonnes are lost from the earth's land surface each year in this way.

Such sediments can be important in the ecology of lakes where the lakes form part of a large river system. This is especially true where large artificial dams have been constructed on major rivers, e.g. Lake Kainji on the River Niger and Lake Nasser on the River Nile. In Lake Kainji, the 'white floods' wash in large amounts of suspended clay into the lake and water transparency is reduced to as low as 10 cm (Obeng, 1981). With the inflow of the clearer waters of the 'black floods', water clarity improves and readings of up to 300 cm are normal. Much of the silt brought into these lakes settles out there, thus decreasing the amounts passing downstream, but rapidly filling in the lake basin. Within 15 years of the completion of the Nasser Dam, the depth of new sediments in some parts of the lake has been estimated at 25m.

Suspended solids can have a major effect on primary production in fresh waters and affects phytoplankton as well as benthic algae and macrophytes. Light penetration can be severely reduced so that

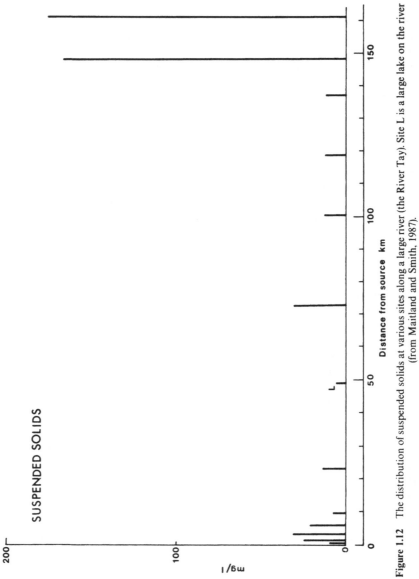

Figure 1.12   The distribution of suspended solids at various sites along a large river (the River Tay). Site L is a large lake on the river (from Maitland and Smith, 1987).

photosynthesis is possible only in the upper layers of the water and silts settling out on the bottom may effectively blanket out algae and rooted macrophytes there. These changes undoubtedly impact on the local fauna, which in addition may be directly affected by the suspended materials. Filter-feeding animals are unable to exist in the presence of large amounts of inorganic suspended solids for their filtering mechanisms simply clog up with inedible material. Many fish too find themselves in difficulties, for not only do they have difficulty in seeing their prey in turbid water, but the particulate material in the water may actually clog their gills.

However, concentrations of suspended solids less than 25 mg/l appear to have little effect on most fish (Alabaster and Lloyd, 1980). Concentrations of inert sediments that do not often exceed 80 mg/l are thought unlikely to do serious damage to a fish population, although they may reduce growth rate and abundance (Hynes, 1973). Thus levels need to remain high (more than 100 mg/l) for several weeks to make a significant impact.

Under natural conditions, very high levels of suspended solids are usually episodic in nature and related to high rainfall and floods in the catchment. Usually the water clears after such an episode and much of the material which has settled on the bottom gradually washes away or is incorporated into the normal substrate. Such changes are responsible for the dynamic nature of many running waters where conditions are normally much less stable than in lakes.

1.2 Chemistry

Almost all natural waters, even rain, contain various chemicals, though the concentrations of different substances may vary greatly from one water to another (Table 1.1). Most waters too (apart from those which are toxic

Table 1.1 Chemical composition of waters in Bohemia draining areas of different geology (after Clarke, 1924). Values in mg/l.

Geology	Ca	Mg	Na	K	HCO_3	SO_4	Cl
Phyllite	6	2	5	2	35	3	5
Granite	8	2	7	4	40	9	4
Mica schist	9	4	8	3	48	10	5
Basalt	69	20	21	11	327	27	6
Cretaceous	134	32	21	16	405	167	17

or of very high temperature) possess living organisms which create a dynamic system, so that the chemical conditions are constantly changing.

Gorham (1961) has pointed out that five different environmental factors interact to determine the ionic concentration of natural waters: climate, geology, topography, biota and time. The origin of the chemicals occurring in fresh waters is not only the soils and rocks of the catchment, but also the atmosphere. Materials in the atmosphere which come from air pollution (especially combustion) by humans and volcanoes, as well as wind-blown sea spray or land dust, enter the system with rain or snow, fall out dry, or are transferred as gas. Substances from the earth, which are normally more important than those from the atmosphere, are dissolved or released by various chemical reactions. The transfer from soil or rock to water is influenced by ion exchange and the type of water involved.

Holden (1966) has shown that the total amounts of salts deposited in rain water in an oceanic area (Scotland) are high compared with continental areas and characterised by a high proportion of cyclic salts. The proportions of sodium, magnesium and chlorides in precipitation were similar to those in sea water, but the proportions of potassium, and especially calcium and sulphate, were in excess. Much of the latter is supposed to be derived from industrial activity.

The concentrations of dissolved solids in natural fresh waters vary greatly, but an average value is approximately 100 mg/l. Water with less than 50 mg/l indicates drainage from igneous rocks, while water with more than 100 mg/l points to drainage from sedimentary rocks. As the amounts of dissolved salts increase, the proportions of calcium and magnesium tend to rise, and those of alkalis to fall. Where large areas of highly organic peaty sediments occur, very soft water with a low concentration of calcium but a relatively high proportion of alkalis may occur.

1.2.1 Oxygen

Oxygen and carbon dioxide are of major importance in aquatic systems, and where variations in concentrations occur they normally show an inverse relationship. The respiration processes of plants and animals and the oxidation processes of breakdown lower the amount of oxygen and increase the carbon dioxide. With sufficient light, the process of photosynthesis by living plants produces the opposite effect—an increase in the amount of oxygen and decrease in carbon dioxide.

A third factor of importance in aquatic systems is the amount of movement and mixing of the water, especially its contact with the atmosphere. This is particularly important with oxygen, but less so with carbon dioxide because of the influence of bicarbonates (see 1.2.2). Diffusion gradients at the surface tend to restore the system to something near equilibrium. If adequate mixing is prevented, e.g. by ice, extreme concentrations may occur, i.e. supersaturation where thick plant growths are exposed to bright light under ice, or anaerobic conditions where there are large numbers of animals or much decomposing organic matter and no light for photosynthesis.

The average atmospheric air is made up of some 210 ml/l of oxygen, 780 ml/l of nitrogen and very small quantities of other gases including carbon dioxide. These gases have, however, different absorption coefficients, the values of which are inversely related to temperature and also affected by pressure. Thus at atmospheric pressure water contains 8.9 ml/l oxygen at 5°C but 6.4 ml/l at 20°C. The effect of pressure means that waters at different altitudes, though saturated with oxygen, may contain different amounts of this gas. The situation is complicated by the fact that with increasing altitude the pressure is lowered and the solubility of oxygen correspondingly reduced, but with increasing altitude temperature is also lowered, and this increases oxygen solubility. At high altitudes there may be low concentrations of oxygen in deep waters because of the small amounts of oxygen in solution at equilibrium. Much higher concentrations would be present in identical lakes at lower altitudes. Ricker (1934a) gives a classical discussion on oxygen solubilities, and there are various nomograms to help in the calculation of saturation values (Figure 1.13).

Temperature is also closely associated with the consumption of oxygen. Respiration and breakdown processes are very dependent on temperature, and their rates of activity may increase by 10% or more with each 1°C rise. There are therefore differences in the oxygen requirements of the same water in summer and in winter, and between similar waters in tropical and temperate areas. These factors influence many of the processes taking place in natural ecosystems, among them organic sedimentation. In warm waters, with adequate oxygen, breakdown processes occur rapidly and there is high mineralisation of sediments; in cold waters the reverse is the case. The oxygen deficit is the difference between the saturation value (at the prevailing water temperature and atmospheric pressure) and the observed value.

The demand for oxygen by oxidisable organic matter has been utilised by limnologists as a means of comparing different waters, especially those

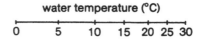

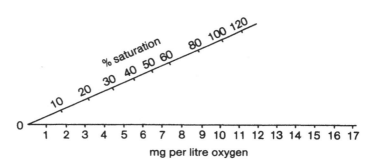

Figure 1.13 A nomogram for obtaining oxygen saturation values at different temperatures. When a line is held to join an observed temperature to a measured oxygen value, it crosses the correct saturation value (after Rawson, 1944).

affected by organic pollution. The biological oxygen demand (BOD) of such waters is a purely arbitrary measure of the amount of oxygen used by a sample of water during a standard period of 5 days at 20°C in darkness. In practice, the water sample is mixed with pure well-oxygenated water, so that about 50% saturation will remain at the end of the test (Table 1.2). The oxygen content of each sample is measured initially and

Table 1.2 A common pollution classification used in Great Britain. The values are in mg/l, and based on a 5-day biological oxygen demand (BOD) at 20°C.

BOD	Classification
0–1	Very clean
1–2.5	Clean
2.5–4	Fairly clean
4–6	Doubtful
6–10	Poor
10–15	Bad
15–20	Very bad
>20	Extremely bad

at the end of the storage period. The test is a useful comparative one, but should be used with caution; its value has been discussed by Phelps (1944) and Klein (1957).

The redox potential (originally called the oxidation reduction potential) is a measure of the oxidising or reducing power of a water. In practice it is the electric potential of one solution relative to another, and it is measured by comparing the electric potential of a bright platinum electrode immersed in a solution containing a mixture of the oxidised and reduced states with that of a standard hydrogen electrode. The redox potential is insensitive to changes in oxygen concentration, and in some natural waters may be the same from the surface (where the oxygen is saturated) to the bottom (where oxygen is virtually depleted). Redox potentials are, however, very sensitive to pH and must be corrected to a definite value (e.g. pH 7).

In many waters, lower potentials occur near the mud/water interface, due to reducing substances released from the mud. Organic sediments themselves are usually strongly reducing and have low potentials, except just at the mud surface under well-oxygenated conditions. This surface layer is usually only a few millimetres thick, and light-brown or yellow in colour, with a redox potential of some 0.5 V. The deeper sediments are usually dark-brown or black with redox potentials as low as 0.0 V. If the amount of oxygen in the water above the mud is depleted, the oxidised zone becomes even thinner and then disappears. The classical work by Mortimer (1941) showed that in one lake a redox potential of 0.6 V occurred just above the mud, but within the mud this dropped rapidly to a minimum at about 5 cm.

1.2.2 Carbon dioxide

Rain falling through the atmosphere absorbs carbon dioxide, and under normal conditions contains about 0.8 mg/l of this gas; the amount may vary from some 1.10 mg/l at 0°C to 0.56 mg/l at 20°C. This dissolved carbon dioxide combines with water to form carbonic acid H_2CO_3. In areas of limestone or other calcium-bearing rocks this acid acts on the rocks to release a soluble product—in the case of limestone, calcium carbonate, $Ca(HCO_3)_2$. At low pH values most of the carbon dioxide is present as such or as undissociated acid. At pH 7 about 20% of the carbon dioxide is present as carbonic acid, and almost all the rest as undissociated HCO_3. At pH values greater than neutral, most of the carbon dioxide is present as HCO_3.

As rain water percolates through soil, it may pick up even more carbon dioxide, especially in areas of limestone or similar rocks. Although relatively insoluble, calcium in these rocks does go into solution as calcium bicarbonate in the presence of carbonic acid. Some free carbon dioxide must be present in the water, however, for this solution to be stable but, if large amounts are available, then high concentrations of bicarbonate may develop.

After passing through the earth, water normally emerges as a spring or trickle of some kind. Calciferous waters then start to lose carbon dioxide to the atmosphere or on mixing with other water; the calcium bicarbonate dissociates, and insoluble calcium carbonate precipitates, often encrusting the substrate and forming marl. This continues until there is sufficient free carbon dioxide to stabilise the solution. The movement of carbon dioxide through the water surface itself appears to be a rather slow process.

If acid water comes into contact with calciferous water, some bicarbonate combines with acid to release free carbon dioxide, which is only weakly dissociated, altering the pH by a small amount. Only small changes in pH can occur until all the bicarbonate is used up. This buffering effect prevents drastic changes in reaction, and is important to aquatic organisms. In most natural waters the buffering ability is determined by the amount of bicarbonate present; this and the pH are useful parameters to measure when investigating their chemistry. Waters which are poorly buffered may exhibit drastic fluctuations in pH and can be alternately acid and alkaline on a circadian or some other basis, depending on local circumstances.

1.2.3   *pH and the hydrogen ion*

Absolutely pure water is weakly dissociated and neutral in reaction, i.e. H^+ and OH^- ions are present in equal proportions. If an acid, base or salt is added to this water, it dissociates and contributes H^+ or OH^- ions, thus altering their concentrations in the solution. Since the dissociation constant of water remains the same, the concentration of the opposite ions alters simultaneously by the same factor (Figure 1.14). Thus data on one ion only are required to define the reaction. The H^+ ion is the simplest, and the one used for this purpose. In pure water at 25°C the concentration of H^+ ions is exactly 10^{-7} moles/l. Conventionally the log of the number of H^+ ions is used, and this is designated pH. When the reaction is acid, the pH is less than 7 (H^+ more than 10^{-7}); when it

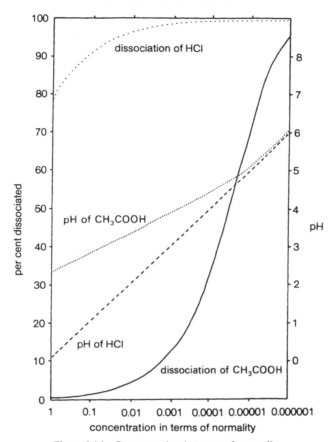

Figure 1.14 Concentration in terms of normality.

is neutral, the pH is 7 (H^+ exactly 10^{-7}); and when alkaline, it is more than 7 (H^+ less than 10^{-7}).

The dissociation constant of pure water increases with temperature; this means that although it has a pH of 7.00 at 25°C, at 0°C the pH is 7.47, at 10°C it is 7.27, while at 40°C it is 6.77. Widely differing values of pH have been recorded from different natural waters, ranging from below 2.0 in certain volcanic lakes which contain free sulphuric acid to more than 12.0 in highly alkaline soda lakes. However, the range of pH found in more normal natural waters is about 4–10.

pH has long been assumed by limnologists to be an important factor in freshwater systems and in many projects it has been assiduously measured. Though a number of studies have indicated some kind of relationship between various groups of organisms and the pH of the waters in which they occur, rarely has any causal association been proved. Many years ago, Hutchinson (1941) doubted that there had been any serious demonstration that pH in itself influences the natural distribution and abundance of aquatic animals, with the exception of a few protozoans. Many recent studies have indicated that there is at least an indirect relationship with many natural plant and animal communities.

With the increasing attention being paid to the phenomenon of acid deposition in recent years, there has been a surge of new interest in pH and its measurement. One result of this is that it has been shown that the accurate measurement of pH is difficult and requires efficient modern equipment with careful attention to regular calibration (United Kingdom Acid Waters Review Group, 1986). This has cast doubt on many of the measurements of pH carried out in the past. There has also been much experimental work on the tolerance of many animals, especially the life stages of various fish, to pH and related factors and much more is now known about this aspect of aquatic ecology.

One of the important aspects of pH and water chemistry which has arisen as a result of recent studies has been the relationship with calcium and with aluminium (some forms of which are extremely toxic to fish), and the subsequent interrelationships with aquatic biota. It has been shown by Chester (1984) that the majority of acid lakes which are fishless are those with a Ca^{2+}/H^+ ratio of less than 3, whereas most of the lakes with healthy fish populations have Ca^{2+}/H^+ ratio greater than 4. Brown and Sadler (1989) have pointed out that solubility of aluminium is a direct function of ambient pH, being at a minimum at around pH 5.5 and increasing towards both ends of the pH scale. However the situation is a complex one because a variety of aluminium species are formed under different conditions and, in addition, aluminium has a strong tendency to form complexes with other anions and with organic molecules.

1.2.4 Nitrogen

Rain water usually contains small amounts of nitrogen compounds, mostly in the form of ammonia and nitric acid dissolved from the atmosphere. As this water percolates through the soil, considerable amounts of nitrogen (originally fixed from the atmosphere by bacteria) may be taken up.

Ammonia is a major breakdown product of plant and animal proteins. In aerobic conditions it is rapidly changed by certain bacteria to nitrate and this, too, may be leached from the soil by rain water. In natural waters, also, many bacteria and some species of blue–green algae are able to fix nitrogen, and can contribute significant amounts to the nitrogen budget of the waters concerned.

Nitrogen is an important component of the cells of living organisms, and the amounts available in a water, though often small, are of significance to the ecosystem. In the majority of natural aerobic waters, most nitrogen occurs as nitrate. In eutrophic waters with large standing crops of algae or macrophytes, almost all the nitrate present may have been incorporated in the plant cells. In anaerobic situations, nitrates are often broken down through nitrite and nitrous oxide, occasionally as far as nitrogen itself. In addition, in the absence of dissolved oxygen, free ammonia may occur as a result of protein breakdown.

1.2.5 Phosphorus

Though an important constituent of living organisms and present in them in significant amounts, the quantities of phosphorus occurring in most waters are low. Occurring naturally mainly as phosphate, the amounts are low, firstly because the element is naturally scarce, and secondly because of the capacity of many plants to absorb and store many times their immediate needs of phosphate. Both phytoplankton and littoral vegetation are capable of doing this; moreover, many of these plants are exceptionally efficient in using phosphorus, and can extract it in significant amounts from water where it is present in minute quantities.

Very little phosphorus is normally present in rain water, but when the rain reaches the earth and percolates through the soil small amounts are leached from phosphorus-containing rocks and from phosphorus present in the soil. The quantities occurring naturally in fresh waters depend on the geochemistry of the catchment, and are greater in areas of sedimentary rocks than in those containing igneous rocks.

In natural waters where phosphorus and iron occur together under aerobic conditions, insoluble ferric phosphate is precipitated into the bottom sediments. In this way, much of the phosphorus may be immobilised in the mud. Where anaerobic conditions occur (during stratification or beneath ice) ferric iron is reduced to the ferrous state, and phosphate is liberated into the water where it is available again as a plant nutrient.

1.2.6 *Iron*

Like nitrogen and phosphorus, iron plays an important part in the metabolism of many organisms, and is also involved in the cycling of certain nutrients (e.g. phosphorus). Iron normally exists in two forms: ferrous, which is soluble but stable only in anaerobic conditions, and ferric, which is stable in the presence of oxygen but is insoluble. Thus in most natural waters there is very little dissolved iron.

Rain water falling on the earth contains some carbon dioxide, but also considerable amounts of oxygen; under these conditions no iron can be dissolved. If, however, as the water percolates through the soil, the oxygen level drops below about 0.5 mg/l and the water comes into contact with ferrous iron, then some of this will go into solution, probably as ferrous bicarbonate. Almost all ferric compounds must be reduced to the ferrous state before they will dissolve, and decomposing organic material may act as a reducing agent to help this. Solution of ferrous compounds is also easier if the pH of the water is low.

When the underground waters which have dissolved large quantities of iron salts emerge as springs or into lakes and rivers where oxygen is available, most of the iron is immediately precipitated as ferric hydroxide $(Fe(OH)_3)$. This is common around some springs and in the bottom sediments of many lakes. Because iron is much more soluble in waters of low pH and can occur in a stable form as a colloidal humate, many brown acid humic waters may contain appreciable amounts of iron even where aerobic. Several bacteria are capable of precipitating ferric hydroxide, e.g. *Siderocapsa coronata* and *Ochrobium tectum*. Iron precipitated in natural waters as ferric hydroxide and forming part of the sediments can readily dissolve again if anaerobic conditions occur (e.g. during stratification).

1.2.7 *Other dissolved solids*

The role of calcium has been mentioned in connection with pH and carbon dioxide. Magnesium, similar in many respects to calcium, is often present in solution as bicarbonate. Its monocarbonate is much more soluble than that of calcium. Silica forms an important part of the skeletal structure of diatomaceous algae and some animals. Occurring normally as undissociated orthosilicate, most is derived from siliceous rocks in the catchment. Sodium and chloride ions occur in most waters, but in low quantities unless the system is affected by saline ground water or ocean spray. Both ions are of major importance in brackish and salt waters. As in all ecosystems, trace elements are important in fresh waters; relatively little, however, is yet known about their true role there.

Dissolved organic matter, once thought to be unimportant in aquatic systems, is now known to be of considerable significance. Birge and Juday (1934), followed by others, demonstrated that dissolved organic substances may occur in far greater quantities than suspended materials in many waters. Important quantitatively as a source of carbon, the quality of the organic material may be of even greater significance, for it may contain valuable amino acids and vitamins. Such materials may originate within a water through the growth and reproduction of organisms; substances derived in this way are termed autochthonous. Materials which originate outside the aquatic system and are subsequently washed or blown in are known as allochthonous.

1.3 Biology

1.3.1 *Decomposition*

The process of decomposition is a fundamental one in nature and can be looked upon both as the end and the beginning of most ecological cycles. The micro-organisms concerned, though all very small, have an enormous variety of form and function and include, as well as bacteria, viruses, streptomyces, yeasts and moulds. Micro-organisms occur everywhere in fresh waters including the surfaces of submerged objects and the particles of bottom sediments. Some can move from the particle surface into water and vice versa when environmental conditions change. In distinct contrast to most plants and animals, micro-organisms are not restricted to a single metabolic category, but can include chemosynthetic, photosynthetic and heterotrophic organisms (Jordan, 1985). Thus microbial production can include primary and secondary production at the same time.

However, the contribution to primary production by the photosynthetic activity of micro-organisms is normally small and restricted to specific aquatic conditions. On the other hand, micro-organisms do make substantial contributions to the productivity of fresh waters by interacting with organic matter originally produced outside the water (allochthonous material) and making it more available to higher organisms. In some systems, the contribution to production from this source can sometimes exceed that from normal plant primary production.

Production by heterotrophic organisms can be of major importance, especially because the micro-organisms concerned can break down organic matter which animals cannot use and produce particulate food from dissolved organic materials. During these processes mineralised nutrients are released for use in conventional primary production by algae and higher plants. The direct value of bacteria as food for zooplankton and other invertebrates is very dependent on the nature and extent of their cell aggregation. Thus, because they are so inter-linked, microbial production cannot easily be separated from microbial decomposition.

Many other important processes are involved during microbial activity in fresh waters. These include nitrification, denitrification, the oxidation of inorganic sulphur compounds and the reduction of sulphates. The numbers of bacteria in lake waters can be enormous and in most lake waters there are between one and ten million bacteria per ml of water.

Microbial production by chemosynthesis occurs in association with anaerobic conditions in water bodies, especially in the boundary layers between aerobic and anaerobic zones. The breakdown of organic materials during anaerobic processes of decay provides reduced inorganic substances which are used by chemoautotrophic bacteria as energy substrates. This can be regarded as a special kind of secondary production and is of special importance in the gradient layers of the redox potential (Sorokin, 1964) though its significance elsewhere is very low.

1.3.2 *Photosynthesis*

Photosynthesis, one of the major life processes occurring on earth, is fundamental as the means of carbon fixation. During this process carbon dioxide is removed from the environment and oxygen is released. Though in total a complex action, photosynthesis is carried out by green plants using energy from solar radiation as follows:

$$6H_2O + 6CO_2\,(\uparrow + 674\,kcal) \xrightleftharpoons{\text{chlorophyll}} C_6H_{12}O_6 + 6O_2$$

Though this reaction is reversible, the reverse (which is respiration) tends to be slower, and normally there is a net gain of organic matter. This and the nutrient salts available are utilised for plant growth.

Although there are usually only small amounts of free carbon dioxide present in solution, in many waters there are considerable reservoirs in the form of bicarbonate. Thus as free carbon dioxide is utilised by green plants, further amounts are released by the bicarbonate, which then precipitates out as insoluble calcium bicarbonate. In many calciferous waters, therefore, high rates of photosynthesis are often accompanied by this precipitation (marl), which encrusts the substrate and often the leaves of the macrophytes themselves.

With most aquatic plants, however (apart from some algae and mosses), the actual system is slightly more complex, for it would seem that these also use HCO_3^- ions from dissociated calcium carbonate and substitute instead OH^- ions. This continues even when all the bicarbonate has broken down, for by hydrolysis the carbonate then gives further HCO_3^- ions, which are replaced by more OH^- ions. The final result is that considerable calcium hydroxide is available, and the water is highly alkaline. This extreme situation occurs only occasionally in nature, usually in small thickly vegetated pools which are exposed to sunlight for long periods.

In most waters the gas exchange system related to photosynthesis is not in constant equilibrium, but usually exhibits a circadian cycle on which a seasonal pattern may be imposed. During darkness, when oxygen is used continuously for respiration by plants, animals and breakdown processes, dissolved oxygen content in the water decreases; it is replaced only by mixing processes at the water surface. During darkness, carbon dioxide too is produced continuously; it is either used to build up a store of dissolved bicarbonate or lost at the water surface. With daylight, however, assuming adequate solar radiation is available, photosynthesis proceeds and oxygen is given off. This may counteract any deficit which has occurred during darkness and, depending on the rate of circulation of the water and the degree of gas exchange at the surface, may even build up supersaturation values.

Because plants need light in order to photosynthesise, the depth to which any species can grow in a water is determined by the extent to which sufficient light for its requirements can penetrate. Different waters may have widely varying light extinction coefficients, depending on the quantity and quality of dissolved and suspended material (including algae) they contain. Generally, there is inadequate light, even in very clear waters, for plants to grow at depths greater than 10 m, and in many eutrophic or turbid waters photosynthesis cannot proceed below about 2 m (Figure 1.15). Even with adequate light the efficiency of its utilisation by plants may be low; Edwards and Owens (1962) have shown that in macrophytes in chalk streams the efficiency of gross daily productivity in relation to solar radiation varies between 1.0 and 2.2%.

1.3.3 Nutrient limitation

As we have seen, a variety of nutrient materials are needed for the growth

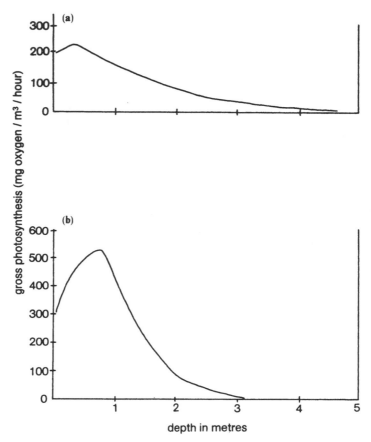

Figure 1.15   Depth profiles of rates of gross photosynthesis of phytoplankton in Loch Leven
(a) in May and (b) in August 1971 (after Bindloss, 1974).

and reproduction of plants in fresh waters. Such nutrients are present in
very variable quantities from one water to another and, even within the
same water system, may vary in space (especially during stratification) or
seasonally according to rainfall, temperature, wind and the nature of the
plant community itself. Usually several nutrients are present in excess of
what might be required for the adequate growth of a particular plant
species. If even one important nutrient is scarce, it may severely hinder
the production of a species and prevent it utilising other nutrients present
in excess (Figure 1.16). Nutrients known to be limiting in natural waters

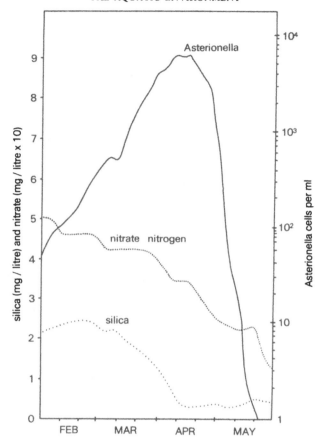

Figure 1.16 Numbers of *Asterionella* cells and concentrations of dissolved silica and nitrate in the 0–5 m water column of Esthwaite Water in 1948 (after Lund, 1950).

are carbon, nitrogen and phosphorus, but the requirements vary with different plant species, and often with each community.

In eutrophic and some tropical waters, nitrogen is a major factor limiting primary production, but in many other waters phosphorus occurs only in minute amounts and is an important controlling factor. Magnesium, potassium, iron, manganese, cobalt, molybdenum and zinc may also be of importance in limiting growth in various waters. Many fresh waters contain a variety of organic compounds which can act as chelating agents. Most organic compounds which algae use as energy sources can also be used by bacteria and fungi. Vitamins are essential for many algae, but their ecological role and importance as limiting factors are still not well known.

1.3.4 *Influence of organisms on the environment*

While it is true that living organisms are profoundly affected by their environment, it is equally important to remember that many organisms are capable of doing the reverse and significantly altering their habitat, sometimes to their own detriment. The influence of the biological component is often relatively greater in freshwater than in marine or terrestrial systems, because of the small size of many waters. There are many examples of the influence of organisms on all kinds of fresh waters, and only a few cases can be considered here.

Many of the important effects of organisms are related to their physiology, especially growth and respiration. Photosynthesis itself is a notable example of a major influence of plants on their environment, and by their subsequent growth many species can deplete essential nutrients, thus limiting their own growth or that of other species. In Windermere the alga *Asterionella* is unable to grow in conditions which it itself has created (Lund, 1950). This plant starts to grow rapidly in the spring in this lake, using up so much of the silica that there is no longer enough to maintain its own growth. The population declines as a result. If an algal population is not limited by nutrients, its rate of growth may be rapid, but, when

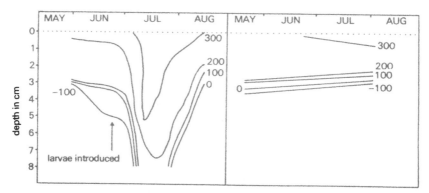

Figure 1.17 Redox potential distribution within settled activated sludge in two experimental tanks. Midge larvae were introduced to the left-hand tank at the point indicated by the arrow. None were placed in the right-hand tank (after Edwards, 1958).

certain densities are attained, self-shading may occur and control the population at this level.

The respiration of animals may use up large amounts of oxygen: Edwards (1958) has shown that the burrowing activities of the larvae of the midge *Chironomus riparius* considerably alter the character of the mud in which they live by extending the redox potential further down into the mud (Figure 1.17). These larvae, when added to natural muds at densities of about $8000/m^2$, raised the oxygen uptake some 25%—equivalent to their predicted respiration. Edwards estimated that natural populations of this species are capable of lowering the oxygen concentration of a stream by almost 1 mg/l per km. Such a process can be taken to extremes where there is no replacement of the oxygen, as in the hypolimnion or under ice in some lakes. Here, oxygen deficiency can be so extreme as to cause massive mortalities among the invertebrates and fish present.

As well as altering the chemistry of their environment, plants and animals often affect its physical characteristics profoundly. The relationship between plants and light extinction has already been discussed. Several species of macrophyte form such large growths in rivers that they seriously affect water levels and velocities. Hillebrand (1950) has shown that in the River Eder, in a stretch where the breadth of the river was some 70 m, the dense weed in summer grew so quickly that the water level rose by over 1 cm per day. Such growths may also change the character of the substrate, for, though originally stony or gravelly, the development of macrophytes impedes water movement and results in increased siltation. In this way the original substrate may be completely obscured by the new sediment. In lakes too the growth of macrophytes may have a profound effect and, through the process of ecological succession, small open waters may be transformed into marshes and eventually into dry land. Completely new lakes, on the other hand, can be created by certain organisms, the dams constructed by beavers being a notable example of this.

CHAPTER TWO

PLANTS AND ANIMALS
OF FRESH WATERS

It is now believed that the living world contains several million different species of organism. These display a fascinating variety of structure and organisation, usually closely adapted to their habitat and mode of life. Before discussing adaptation, community structure and energy cycles in fresh waters, it is useful to consider the range of plants and animals which occurs there. The survey here, though general, is fairly comprehensive. However, it is biased towards freshwater organisms (Figure 2.1), and several major groups are not included at all because they are absent from fresh water, e.g. the Echinodermata, Sipunculida, Hemichordata, Pogonophora, Chaetognatha and several other phyla are exclusively marine, while the Onychophora and many groups of Arachnida are entirely terrestrial.

The systematic classification adopted follows those of McLean and Ivimey-Cook (1956) and Rothschild (1961)—apart from Chordata, which follows Young (1955). The major divisions are equivalent to a sub-kingdom or phylum, the minor groupings to a class or sub-class. The intention has been to provide a brief account of the main characteristics of each group of organisms, its form or feeding habits (where relevant) and its importance in freshwater habitats throughout the world. These accounts are minimal; further appropriate references are found in Smith (1951), McLean and Ivimey-Cook (1956), Kimball (1965), Barnes (1968) and Russell-Hunter (1968, 1969).

2.1 Viruses

Viruses (Figure 2.2(1)) are all extremely small, ranging in size from 30 to 300 μm, and cannot be seen with the conventional light microscope. They are radically different from all other members of the organic kingdom,

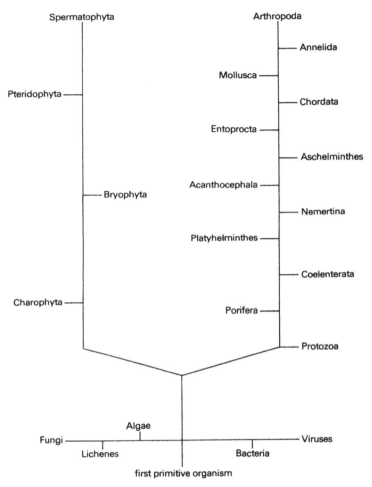

Figure 2.1 A diagrammatic representation of suspected evolutionary relationships among groups of freshwater organisms.

but they can undoubtedly reproduce themselves and evolve, changing their properties over a number of generations. All viruses are obligate parasites, being able to grow only within the cells of other living organisms. Their occurrence is world-wide and in all environments, where they attack a great variety of hosts, both plant and animal.

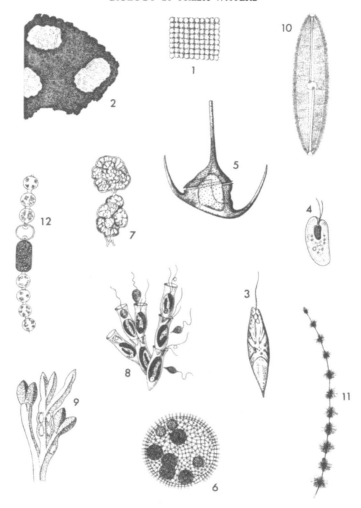

Figure 2.2 (1) Virus (tobacco necrosis), (2) bacteria (*Siderocapsa*), (3) Euglenophyceae (*Euglena*), (4) Cryptophyceae (*Cryptomonas*), (5) Dinophyceae (*Ceratium*), (6) Chlorophyceae (*Volvox*), (7) Xanthophyceae (*Botryococcus*), (8) Chrysophyceae (*Dinobryon*), (9) Phaeophyceae (*Fucus*), (10) Bacillariophyceae (*Navicula*), (11) Rhodophyceae (*Batrachospermum*), (12) Cyanophyceae (*Anabaena*).

2.2 Bacteria

Bacteria (Figure 2.2(2)) are also very small (about 1–50 μm long) and mostly unicellular. Though they resemble the Cyanophyceae (see 2.3.10), only a few are photosynthetic. Bacteria have a rigid cell wall but not a distinct nucleus. Most species are sedentary, but a few can glide or swim by means of flagella. Three shapes of cell are found—rod-like (bacilli), spherical (cocci) and curved (spirilli). Unlike most other organisms, bacteria are classified not only on morphology, but also on physiology and reactions to various tests. Members of the 2000 or more species are found in all parts of the world in all environments as free-living forms, and inside many plants and animals. Bacteria are responsible for many plant and animal diseases, including some fatal to humans. They are also of major importance in all habitats in breaking down dead plants and animals, and in recycling and releasing chemicals to the organic cycle.

2.3 Algae

Algae are of major importance in aquatic communities, and each class is worth considering separately. Tiffany (in Smith, 1951) notes that no other group, except bacteria, is able to grow in such diverse conditions. The ecological relationships among freshwater algae are varied and complex. They occur in all fresh waters, as well as in the sea and in some terrestrial habitats. Most forms are planktonic; others are associated with a substrate as benthophytes, epiphytes or even epizoophytes. Some have developed symbiotic relationships with certain fungi to form lichens.

2.3.1 Euglenophytes

These flagellate algae are green or colourless (Figure 2.2(3)) with two flagella (one very short) originating in an anterior gullet. Chloroplasts, when present, are green. Euglenophytes are common in marine, brackish or freshwater pools with an abundance of organic matter. Some occur in mud bordering aquatic habitats, others in soil, while some sessile species grow on other algae, plant debris or invertebrates; many are commensal with animals. In rich fresh waters, some species occur in sufficient quantities to colour the water. Some are phototrophic and utilise carbon dioxide in the presence of light; others are heterotrophic and have no chlorophyll. These forms depend on organic compounds as a source of carbon.

2.3.2 Cryptophytes

This is a small group of flagellate algae (Figure 2.2(4)) occurring in marine and freshwater systems. Some have no gullet, and laterally inserted flagella; others have a longitudinal gullet with two flagella anteriorly. There may be a few large (or many small) chromatophores which vary in colour from yellow–green to brown or red. Some forms are colourless.

2.3.3 Dinophytes

The algae in this group (Figure 2.2(5)) occur both in the sea and in fresh water; some species are symbiotic in invertebrates or fish. All Dinophyceae have a very characteristic bead-like nucleus. Some are colourless, but yellow–green or yellow–brown forms are the most common. All species have two flagella, which often lie within the grooves encircling the body in various ways. All types of algal nutrition are found within this group. Of the three main orders, one is entirely marine (Dinophysiales) but the two others (Gymnodiniales and Peridiniales) contain both marine and freshwater representatives.

2.3.4 Chlorophytes

Members of this important group, the green algae (Figure 2.2(6)), do occur in the sea and in brackish water, but most are found in fresh waters—of widely ranging types, including semi-aquatic habitats such as moist stones, trees and soil. Many are planktonic, others attach to the substrate, are epiphytic, or even float about. Some are symbionts, others true parasites. As expected from this wide range of habitat, the Chlorophyceae exhibit a great variety of forms, all of which possess green chloroplasts.

2.3.5 Xanthophytes

These algae (Figure 2.2(7)) include forms found in both marine and freshwater habitats. They are characterised by the possession of numerous, often discoidal, yellow–green chromatophores and oil globules. Xanthophyceae exhibit a range of vegetative structure—some are filamentous, others siphoneous or coccoid—but most possess (at least during one life history phase) two unequal anterior flagella.

2.3.6 *Chrysophytes*

These are characteristically golden-yellow or brown algae (Figure 2.2(8)) due to accessory carotenoid pigments in the otherwise green chromatophores. The uninucleate cells have large plate-shaped chromatophores, and silicified cysts are characteristic; these are formed during unfavourable conditions. The motile Chrysophyceae exhibit a greater diversity of form than any other algal class except possibly the Chlorophyceae. Some are unicellular and naked, or with a protective sheath; many others are colonial. A few are epiphytic. They may be found in marine, brackish or freshwater habitats. The numerous freshwater species are most successful in cool habitats or during the cool period of the year.

2.3.7 *Phaeophytes*

Commonly known as brown algae (Figure 2.2(9)), these owe their olive-green to dark-brown colour to an accessory carotenoid pigment (fucoxanthin) in the chromatophores. This masks other pigments there. The Phaeophyceae are almost entirely marine, though there are brackish water forms and three genera (*Bodanella, Pleurocladia* and *Heribaudiella*) in fresh water. Most grow as attached forms in the sublittoral intertidal zone, though a few float freely—most notable among which is *Sargassum*, which grows as enormous masses in the Sargasso Sea.

2.3.8 *Bacillariophytes*

The diatoms, as this class is commonly known, have a characteristic cell membrane which is silicified and of complex structure (Figure 2.2(10)). The ornamental cell wall has two halves (valves), one overlapping the other. There are two main orders. In the Centrales the valves are circular with concentric markings; in the Pennales the valves are elliptical with markings in two series, one on either side of the median line. Some diatoms are able to glide by means of an axial structure known as a raphe. Many diatoms are unicellular, others are colonial, forming long filamentous colonies. They occur in the sea, fresh waters and the soil.

2.3.9 *Rhodophytes*

These algae (Figure 2.2(11)) are characterised by complex thalli which are reddish in colour. Commonly known as red algae, they are uniform as a group and exhibit a distinctive type of reproduction. Almost all are marine and found as attached forms in the lower intertidal and sublittoral zones (often at great depths), especially in warmer seas. From a total of some 4000 species, only about 200 occur in fresh water.

2.3.10   *Cyanophytes*

The Cyanophyceae (Figure 2.2(12)), known as blue–green algae, have cells resembling those of bacteria. These cells, surrounded by a membrane, have no organised nuclei or central vacuoles, each being a mass of protoplasm, within which are pigments and granules. Most cells are cylindrical or spherical and, though some species are unicellular, the majority form multicellular colonies. Blue–green algae occur in most habitats throughout the world. The greatest growths of many species are found in sunny, rich, shallow waters. Some species are capable of fixing nitrogen.

2.4   Fungi

Fungi (Figure 2.3(1)) are characterised by having no chlorophyll and living as saprophytes on decaying organic matter, or as parasites on other organisms. The vegetative part consists of a complex of filaments (hyphae) which make up the main body (mycelium); in saprophytes this grows among the organic matter, and on or in the host in the case of parasites. The reproductive bodies develop as spores which may be borne on small and insignificant organs, or on large and conspicuous growths (e.g. mushrooms). The majority of fungi are terrestrial (mainly in damp situations), but there is a large number of aquatic species, mainly in fresh waters. Of the four main classes, only the Phycomycetes are predominantly aquatic; within this class the order Chytridiales is mainly ectoparasitic on algae, while the Saprolegniales occur on decaying plants or animals in fresh water and even on the injured tissues of living organisms. The Monoblepharidales are saprophytic, living submerged in fresh water on decaying twigs.

2.5   Lichens

A lichen (Figure 2.3(2)) is never, in fact, a single species or organism, but is always composed of both fungal and algal cells (known respectively as

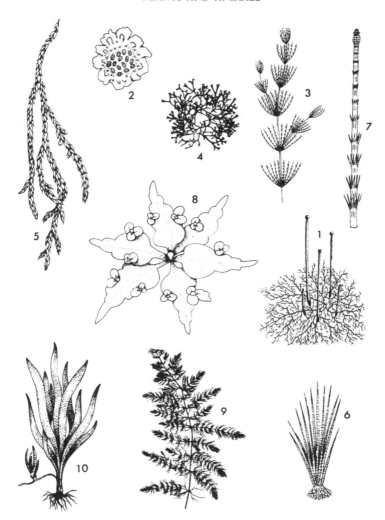

Figure 2.3 (1) Fungi (*Mucor*), (2) Lichenes (*Xanthoria*), (3) Charophyta (*Chara*), (4) Hepaticae (*Riccia*), (5) Musci (*Fontinalis*), (6) Lycopsida (*Isoetes*), (7) Sphenopsida (*Equisetum*), (8) Pteropsida (*Ceratopteris*), (9) Dicotyledones (*Ceratophyllum*), (10) Monocotyledones (*Sagittaria*).

hyphae and gonidia) associated in symbiosis. The vegetative portion (thallus) of the lichen has one of three growth forms: fruticose (strap-like), foliose (leaf-like) or encrusting. In most cases the fungus is an ascomycete, though in a few tropical forms basidiomycetes are concerned; these fungi are found only in the lichens. The algae are unicellular Chlorophyceae or Myxophyceae, but many of the species occur independently. Lichens occur in most parts of the world, and in the Arctic and Antarctic may be the dominant vegetation. The majority of lichens are terrestrial, many occurring on rocks along sea coasts and on mountains. A number occur in shallow fresh waters, notably on submerged rocks in mountain streams.

2.6 Stoneworts

A small group of aquatic non-flowering plants with a single order (Charales), the stoneworts are related to algae, but do possess unique characteristics. They have no proper roots, but are anchored by root-like growths called rhizoids. The stems are jointed and branching (Figure 2.3(3)); the large single cells contain chlorophyll and each makes up a section of the stem. In calcareous waters the stems become encrusted by lime, and have a stony texture which gives the common name. Stoneworts occur in still or slow-flowing fresh water, but occasionally in brackish situations. The two most widespread genera are *Chara* and *Nitella*.

2.7 Bryophytes

In the Bryophyta, the body is of a simple structure, consisting of a flat, elongate thallus. The reproductive organs are complex, and there is often an alternation of generations in the form of the plant.

2.7.1 *Liverworts*

Commonly known as liverworts (Figure 2.3(4)), Hepaticae are always dorsiventral and frequently prostrate and thalloid. The reproductive spores are developed in sporangia. In lower liverworts the thallus is prostrate, flat and branching and secured to the ground by simple rhizoids. In higher forms an axis develops and leaves are produced on it. Most liverworts are terrestrial, but grow in shady moist places, usually near streams or on marshy ground; a few species occur in fresh water (e.g. *Jungermannia* and *Riccia*).

2.7.2 *Mosses*

The Musci (Figure 2.3(5)), or mosses, differ from the Hepaticae in having clearly defined stems and leaves. There is, however, no development of true vascular tissues as in higher plants. Reproductive spores develop in a specialised capsule on a stalk. Like liverworts, most mosses are terrestrial and grow in damp, often shady, places—though some species can withstand desiccation. A few mosses occur in fresh water, e.g. *Fontinalis*, which is common in shaded streams, and *Sphagnum*, often abundant in peat pools. In such systems, they may be the only larger plants present.

2.8 Pteridophytes

This group includes all the vascular non-flowering plants. Unlike the Bryophyta they have well-developed roots; the aerial parts of each plant are divided into stems and leaves. The Pteridophyta produce spores which give rise to the small gametophyte stage. The group varies greatly in form and is mainly terrestrial—though each class has aquatic representatives.

2.8.1 *Lycopsids*

About 1000 species occur in this group (Figure 2.3(6)), and many have a superficial resemblance to mosses in growing close to the substrate, possessing small leaves and bearing spores in club-like structures called strobili. The Lycopsida are, however, true vascular plants with conducting vessels in both root and stem. Of the two orders, the Lycopodiales (club-mosses) are almost entirely terrestrial, while the Isoetales (quill-worts) are largely aquatic, occurring in various aquatic habitats around the world (mainly at high altitudes in the tropics). There are no marine species. Within the family Isoetaceae are two genera: *Isoetes*, a cosmopolitan genus, and *Stylites*, known only from lakes in South America.

2.8.2 *Horsetails*

There are only some 25 living species in this group, commonly known as horsetails because of the characteristic whorls of branchlets arising at

regular intervals from the main stem (Figure 2.3(7)). All living species of
horsetails belong to one order (Equesitales) and a single genus *Equisetum*.
This has freshwater and terrestrial species in different parts of the world.

2.8.3 Ferns

Pteropsida (ferns) differ from other Pteridophyta in having large
leaves, with branching veins (Figure 2.3(8)). The vascular roots and
stems grow underground, the leaves (fronds) growing up from the rhizome
each spring. Sporangia on the fronds carry the reproductive spores. Ferns
occur in freshwater and terrestrial habitats in many parts of the world,
but are never marine. Of the four main freshwater families only the tropical
Ceratopteridaceae are really fern-like. The other three (Marsileaceae,
Azollaceae and Salviniaceae) are specialised and reduced in vegetative
structure, being mostly small, free-floating forms found in still, fresh waters.

2.9 Spermatophytes

The Spermatophyta is a large group characterised by their method of
reproduction and distribution—the seed. This has an outer covering, the
testa, which encloses the embryo and its reserve food material, the
endosperm. In many higher members, flowers and fruit have evolved, and
the Spermatophyta have been successful not only in terrestrial habitats,
but also in many fresh waters and even a few marine localities.

2.9.1 Dicotyledons

In this group of Spermatophyta (Figure 2.3(9)) the embryos possess two
cotyledons on germination. The leaves of most Dicotyledones are
net-veined and have petioles. Internally the vascular bundles are arranged
around a pith. This large group, with more than 90 000 species, occurs in
all parts of the world, in both terrestrial and freshwater habitats. Though
some tolerate brackish conditions, there are no marine representatives.
Sculthorpe (1967) notes that ten families are mainly aquatic; most of these
are cosmopolitan (viz. Nymphaeaceae, Ceratophyllaceae, Elatinaceae,
Haloragaceae, Callitrichaceae and Menyanthaceae) but one (Hippuri-
daceae) is found only in north temperate and cool South American areas,
while the others (Trapaceae, Podostemaceae and Hydrostachyaceae) are
mainly tropical. Most of these families occur in standing waters or in fine

substrates in running waters. The Podostemaceae, however, are unique in growing among rocks in torrents. Aquatic species are also found in families which are otherwise terrestrial; most are emergent forms (e.g. *Polygonum*: Polygonaceae), though some are floating (e.g. *Pistia*: Araceae) or entirely submerged (e.g. *Hottonia*: Primulaceae).

2.9.2 *Monocotyledons*

In this smaller group of Spermatophyta (Figure 2.3(10)) only one cotyledon is produced by the embryo at germination. In mature plants, ovate or linear leaves are typically parallel, veined with a sheathing base. The primary root rarely persists beyond the seedling stage, being replaced by adventitious roots from the stem. In the stem vascular bundles are scattered. The Monocotyledones—with some 50 000 species—are found in terrestrial and freshwater habitats in all parts of the world. A few marine species inhabit coastal waters in tropical seas, though some (e.g. *Zostera*: Zosteraceae) are temperate. Some eighteen families are almost exclusively aquatic; many are cosmopolitan (e.g. Hydrocharitaceae, Alismaceae, Potamogetonaceae, Zannichellaceae), but some occur only in north temperate areas (Scheuchzeriaceae) or in the tropics (Aponogetonaceae). As well as these exclusively aquatic families, hydrophytes occur in what are otherwise terrestrial families of Monocotyledones (e.g. *Glyceria*: Graminaceae; *Crinum*: Amaryllidaceae).

2.10 Protozoans

Protozoa are diverse unicellular microscopic non-green organisms. Most are solitary, but there are some colonial forms. Though they have no organs or tissues, each cell can carry out the basic physiological processes, and some species have a complex intracellular structure. Protozoa are primarily aquatic, occurring wherever water is present—in the soil, the sea and all fresh waters. All types of feeding are found in the group and, though most forms are free-living, some are commensal, symbiotic or parasitic.

2.10.1 *Zoomastigines*

Most of this class possess flagella as adult locomotory organs. They have no chloroplasts. Most Zoomastigina (Figure 2.4(1)) have a distinct

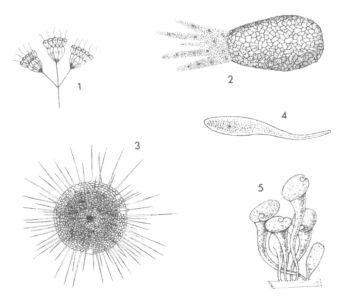

Figure 2.4 (1) Zoomastigina (*Codosiga*), (2) Rhizopoda (*Difflugia*), (3) Actinopoda
(*Actinosphaerium*), (4) Sporozoa (*Monocystis*), (5) Ciliata (*Stentor*).

anterior end: even colonial forms. The one or more flagella provide
movement on the same principle as an aeroplane propeller; some species
have a thin flexible outer pellicle and are capable of amoeboid movement.
Free-living species are holozoic in their feeding, but others are commensal,
symbiotic or parasitic, especially on Arthropoda and Vertebrata. Parasitic
species have complex life cycles involving more than one host; free-living
forms are found in both marine and freshwater habitats.

2.10.2 *Rhizopods*

Adult stages in this class (Figure 2.4(2)) have characteristic flowing
projections known as pseudopodia; these are used for capturing prey and
locomotion. Rhizopoda exhibit a variety of form among four orders. The
Rhizomastigina are asymmetrical amoeboid forms with one flagellum, not
possessed by the Amoebina, which are also naked and asymmetrical, with
a constantly changing shape. The Testacea are symmetrical with an
external shell in which there is an opening through which pseudopodia
can be extended. In the Foraminifera, pseudopodia are fine and branch

frequently, forming a network; most species have a calcareous chambered skeleton of some kind. Apart from some parasitic species, this class is holozoic and feeds on other small organisms. The Rhizomastigina include many marine, freshwater and terrestrial forms, as do the Amoebina, which may also be parasitic. The Testacea are primarily freshwater, while most Foraminifera are marine.

2.10.3 Actinopods

This group of Protozoa (Figure 2.4(3)) all have a spherical body divided into an inner part, with one or more nuclei, and an outer part, from the surface of which radiate needle-like pseudopodia supported by axial rods. There is often a siliceous skeleton. Most Actinopoda are planktonic— though a few are sessile—and feed mainly on living algae and other protozoans, engulfed by the pseudopodia. Of the two main orders, the Radiolaria are entirely marine, but the Heliozoa are mainly freshwater.

2.10.4 Sporozoans

The class (Figure 2.4(4)) is entirely parasitic, with a variable body shape, usually oval or round, with few locomotory organs in the adults, though flagella or pseudopodia may occur during development. Many immature stages are spore-like, with a resistant covering during the transmission phase. All Sporozoa are endoparasites of aquatic and terrestrial animals. Food is absorbed directly through the body surface. Life histories are frequently complex and involve an alternation of sexual and asexual reproduction between generations.

2.10.5 Ciliates

This is the largest class of Protozoa and includes more than 6000 species all with distinctive cilia (Figure 2.4(5)). Two types of nuclei are present, and there is a well-defined mouth. Body shape, though in most cases asymmetrical, is normally constant. A few species are sessile and colonial, but most are free-swimming and solitary. Apart from some parasitic forms, most Ciliata are holozoic and feed raptorially on other protozoans or rotifers, or by cilia which drive food particles (suspended organic matter and bacteria) towards the mouth. Free-living Ciliata are widely distributed in both marine and freshwater habitats. There are few parasites, but many commensal and some symbiotic forms.

2.11 Sponges

Sponges (Porifera) are the most primitive of multicellular animals and possess neither true tissues nor organs. All sponges are sessile (Figure 2.5(1)), their only movement being in the generation of currents created by flagella on cells known as choanocytes. These line the inner cavities of the sponge. The outer surface is perforated by holes called ostia, through which the incoming currents pass. The skeleton is distinctive and often complex, made of siliceous or calcareous spicules, or of spongin fibres. Sponges feed on fine organic particles, including small plankton, drawn in by inhalent currents. All sponges are aquatic but, though there are some 5000 species, these are exclusively marine except for a single freshwater family (Spongillidae). These are widespread in a variety of fresh waters, mainly as encrusting (Figure 2.6) or branching forms. Some are green due to the presence of Zoochlorellae.

2.12 Coelenterates

The Coelenterata (jellyfish, sea anemones, etc.) are radially symmetrical and have tentacles bearing unique cnidoblast cells (Figure 2.5(2)) which enclose stinging nematocysts. These can be discharged and are used for anchorage, defence and the capture of food. Two main body forms occur within the phylum: (a) sessile polyps with cylindrical bodies attached to the substrate at one end and bearing tentacles surrounding the mouth/anus at the other; (b) free-swimming medusae, which are saucer-shaped with a fringe of tentacles. All coelenterates are carnivorous, feeding on other animals which are paralysed and entangled by nematocysts discharged from the tentacles. Of the 9000 living species, all are marine except a few Hydrozoa which are freshwater. Common genera include the polypoid solitary *Hydra* and colonial *Cordylophora*, and the medusoid *Craspedacusta*.

2.13 Platyhelminths

These animals are bilaterally symmetrical and dorsoventrally flattened. They are elongate, oval or ribbon-like in form and the alimentary canal has only one opening, which serves as both mouth and anus. Most free-living forms are aquatic and occur in a variety of freshwater, marine and

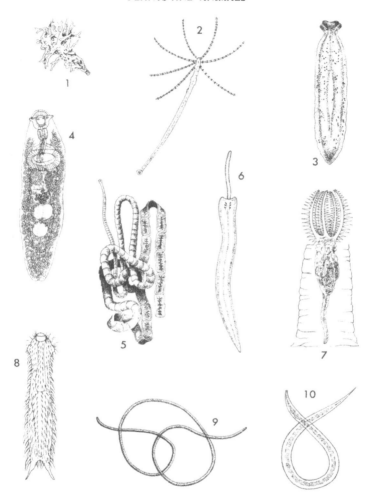

Figure 2.5 (1) Porifera (*Spongilla*), (2) Coelenterata (*Chlorohydra*), (3) Turbellaria (*Bdellocephala*), (4) Trematoda (*Crepidostomum*), (5) Cestoda (*Taenia*), (6) Nemertina (*Prostoma*), (7) Rotifera (*Stephanoceros*), (8) Gastrotricha (*Chaetonotus*), (9) Nematomorpha (*Paragordius*), (10) Nematoda (*Dolichodorus*).

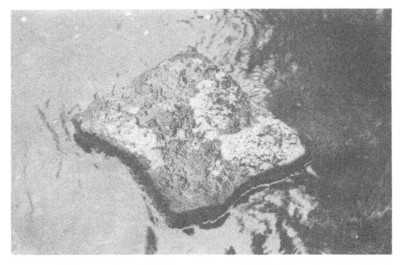

Figure 2.6 Whitish growths of a freshwater sponge (*Ephydatia*) on the underside of an upturned stream boulder (Photo: P.S. Maitland).

terrestrial habitats. There is a large number of parasitic species—two out of the three classes being almost exclusively so.

2.13.1 *Flatworms*

Commonly known as flatworms, Turbellaria are elongate ovoid and dorsoventrally flattened (Figure 2.5(3)). The head bears eyespots and a pair of tentacles or lateral projections. Most species are dark in colour, though a few are brightly coloured or green (from symbiotic algae). Flatworms are carnivorous, feeding on small invertebrates (often dead or dying ones) which are engulfed by the pharynx. Undigested material is egested through the pharynx also, as there is no anus. Most Turbellaria are marine, though there are some freshwater and a few terrestrial forms. Freshwater species are entirely benthic and occur in standing and running waters, often in large numbers where conditions are suitable.

2.13.2 *Flukes*

This entire class (Trematoda, trematodes) are parasitic for most of their lives. The Monogenea are parasitic on a single host, usually a fish or amphibian; the Digenea are endoparasitic on two or more hosts, usually an invertebrate followed by a vertebrate. Trematoda are colourless, elongate ovoid in shape and dorsoventrally flattened (Figure 2.5(4)) with two or more suckers, the anterior one surrounding the mouth. Though some species occur in terrestrial vertebrates, almost all are confined to aquatic (both marine and freshwater) hosts for part of their life history.

2.13.3 Tapeworms

Cestoda, commonly known as tapeworms, are also entirely parasitic, usually in vertebrates as adults, though the intermediate stages may occur in invertebrates. Unlike the Trematoda, many species occur in terrestrial animals and do not need an aquatic state. The body is very long and narrow and made up of numerous segments called proglottids (Figure 2.5(5)); these fill with eggs near the posterior end of the body, where they drop off and are dispersed. The characteristic head has several suckers for attachment to the host's gut wall; there is no mouth, food material being absorbed through the body surface.

2.14 Proboscis worms

These extremely elongate, usually ribbon-shaped, animals (Nemertina, nemertine worms) may reach a length of 1 m or longer (Figure 2.5(6)). The head may be spatulate or lanceolate, and at its anterior end opens a proboscis pore, through which the proboscis can be rapidly everted. Locomotion is by cilia on the body over a path of secreted mucus. A few species can swim. All proboscis worms are carnivorous and feed on Annelida as well as small Mollusca and Crustacea. Most types are benthic and, of the 500 or more species, all are marine except for a few freshwater genera and one terrestrial genus.

2.15 Aschelminths

These are cylindrical elongate animals with an anterior mouth and sense organs, but no definite head. The body is enclosed in a well-developed cuticle, and adhesive glands are common. An unusual feature of most Aschelminths is that the number of cells making up each organ is usually constant within any species. Most forms are small and aquatic.

2.15.1 *Rotifers*

Though they may be as long as 2 mm, most rotifers are small and among the tiniest of the metazoan animals. The body is enclosed in a sculptured cuticle, and the anterior end carries a ciliated organ (corona) which may be modified in various ways (Figure 2.5(7)). The name of the group derives from the rotating appearance of the corona when the cilia are beating. The trunk bears three projections called toes and is usually transparent, though it may appear to be pigmented from coloured material in the gut. Most rotifers are freshwater, though a few marine and semi-terrestrial forms exist. Many are ciliary feeders but there are carnivorous forms feeding on other rotifers or protozoans. Several species build a protective tube. Most rotifers are cosmopolitan, inhabiting various fresh waters, attached to vegetation and stones, or as members of the plankton. Semi-terrestrial species are associated with lichens or mosses, and are active only when these plants have an adequate water film over them (from rain or spray); during dry periods they form cysts which can survive desiccation. Some aquatic species are epizoic and a few are parasitic.

2.15.2 *Gastrotrichs*

These are small metazoan animals about the same size as rotifers. The head region is distinct from the body, which is elongate and convex dorsally but with a flattened ventral surface bearing cilia; these are also found on the head and used for locomotion. The posterior end of the trunk is often forked, while the whole body is enclosed in a cuticle of scale-like parts which may adjoin or overlap and carry long spines (Figure 2.5(8)). Most species have adhesive organs. Gastrotricha are common in aquatic habitats; though one order is restricted to the sea, the Chaetonotoidea are mostly freshwater with only a few marine species. The food is small particles (bacteria, algae and protozoans) sucked into the mouth by the pumping pharynx.

2.15.3 *Hairworms*

Only adult nematomorphs, known commonly as hairworms, are free-living, the young being parasitic in Arthropoda. The body is long—often more than 30 cm—but very narrow and usually dark in colour (Figure 2.5(9)). The exterior cuticle is thick and the gut is rudimentary, for adults do not feed. The order Nectonematoidea is marine, but all members of the only other order, Gordiodea, live in fresh water or damp soil. They are found in fresh waters in most parts of the world. Eggs laid in the water hatch into larvae, each with a spinous proboscis; these penetrate arthropod hosts where development is completed, and the adults leave these hosts when they are in or near water.

2.15.4 *Roundworms*

Nematodes, commonly known as roundworms, may reach a length of several centimetres, but most species are less than 1 mm in length. The body is cylindrical and extremely elongate with tapered ends (Figure 2.5(10)). Many are parasites, but there are also numerous free-living species, many of which are carnivorous, feeding on other nematodes; the herbivorous species eat algae or suck the tissues of higher plants. Some species in decaying organic matter feed on bacteria. The class includes some of the most widely distributed and abundant of all animals, free-living forms being found in all parts of the world. Freshwater forms are normally benthic in habit and occur in many fresh waters, including fast-flowing streams, where anchorage is obtained by means of caudal adhesive glands.

2.16 Acanthocephalans

All acanthocephalans (Figure 2.7(1)) are parasitic and require two hosts for completion of their life cycle. The young stages occur in arthropods; adults are found in the guts of various vertebrates, notably (in fresh water) fish. The body is elongate, and anteriorly the proboscis (for attachment within the host) has many recurved spines; there may also be spines on the body. There are only about 500 species, but some occur in considerable numbers in vertebrate hosts and do considerable damage to the gut wall.

2.17 Entoprocts

The Entoprocta are small animals, always less than 5 mm in length, which may be solitary but are more often colonial in habit (Figure 2.7(2)). All species are sedentary and live attached to rocks, or are epizoic on large crustaceans or some other invertebrates. Apart from *Urnatella*, all

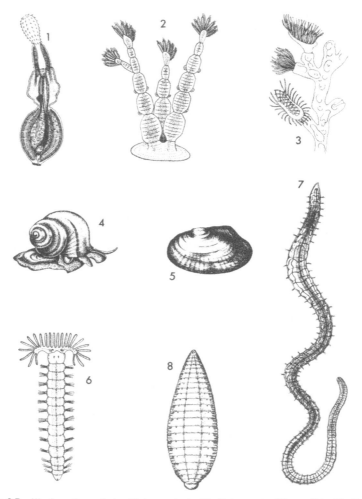

Figure 2.7 (1) Acanthocephala (*Polymorphus*), (2) Entoprocta (*Urnatella*), (3) Polyzoa (*Plumatella*), (4) Gastropoda (*Viviparus*), (5) Bivalvia (*Anodonta*), (6) Polychaeta (*Manayunkia*), (7) Oligochaeta (*Tubifex*), (8) Hirudinea (*Glossiphonia*).

entoprocts are marine. The body consists of an ovoid calyx attached at one end to the substrate by a stalk; the other end bears a crown of tentacles enclosing both mouth and anus. Entoprocts are ciliary feeders on fine organic matter, including small phytoplankton and zooplankton. Cilia on the tentacles transport food along grooves into the mouth.

2.18 Polyzoans

Most polyzoans (Figure 2.7(3)) are sessile and colonial. Each individual is enclosed in a characteristic outer zooecium, which is normally chitinous but may be gelatinous. The animal is attached inside the zooecium, and a tentacle sheath encloses the lophophore, a horseshoe-shaped organ bearing some 15–100 ciliated tentacles. The mouth opens within the lophophore, while the anus opens outside it. Polyzoans feed on organic matter, including small plankton passed to the mouth by cilia on the lophophore. Of the two classes, the Gymnolaemata is almost exclusively marine (the exceptions being *Paludicella* and a few other genera) while the Phylactolaemata is restricted to fresh waters. The latter class, though very widespread, includes only about 50 species; some genera (*Plumatella* and *Fredericella*) are cosmopolitan. Polyzoa occur in a wide variety of fresh waters, excluding those which have insufficient plankton or excessive silting.

2.19 Molluscs

This group includes many of the largest types of invertebrate and contains more species—about 80 000—than any other phylum except the Arthropoda. Mollusca are basically bilaterally symmetrical, and possess a protective shell which is usually made up of calcium carbonate secreted by the mantle. Locomotion is by means of a muscular foot drawn into the shell by powerful retractor muscles. Most molluscs are marine (the classes Amphineura, Scaphopoda and Cephalopoda are restricted to this environment) but many species occur in freshwater habitats; some are terrestrial.

2.19.1 *Snails*

The Gastropoda (Figure 2.7(4)), mainly various snails, contains more than 35 000 living species. Typically each is enclosed in a shell which has the form of a spire. The foot is a flattened creeping sole used for locomotion or attachment. The foot in many groups has a hard operculum which completely fills the shell opening after the animal has withdrawn, acting as a protective door. Some are ciliary feeders, herbivores, carnivores and parasites, but most freshwater forms are herbivorous, feeding by means of a radula. This is tongue-like with numerous rows of rasping teeth; it may act as a rasp, brush, grater or comb, tearing off pieces of macrophytes or collecting organic debris or algae and transporting them to the mouth. There are two main sub-classes: the Prosobranchia, which have gills and are mainly marine forms (a few are freshwater), and the Pulmonata, which have no true gills and are mainly freshwater or terrestrial with only a few marine species.

2.19.2   Bivalves

All bivalves (Figure 2.7(5)) are laterally compressed, and the shell is made up of two valves which are hinged dorsally and pulled together by two large adductor muscles. The foot is large and used for burrowing and locomotion, while the head is small. The mantle cavity within the shell is large, as are the gills lining it. Most bivalves are ciliary feeders on small organic particles swept into the mantle cavity by currents created by cilia on the gills. These also trap food (by secreted mucus) and pass it along grooves to the mouth. A few species feed on detritus by means of a muscular proboscis. All bivalves are aquatic, and most are marine; a number do occur in fresh water, however, and can be important in certain habitats. Most occur in soft substrates, but some (e.g. *Dreissenia*) attach themselves to hard substrates.

2.20   Annelid worms

Annelids are elongate worm-like animals in which the body is divided into numerous segments arranged in linear series. Many body organs (nerves, muscles, excretory organs) are repeated in each segment, though the gut passes through all segments. These animals occur in a wide variety of habitats and are of major importance in many communities.

2.20.1   Polychaetes

Polychaete worms have segmental pairs of lateral appendages known as parapodia, each with numerous stiff bristles (Figure 2.7(6)); they are used for locomotion and respiration. Though many species live in burrows, others are free-swimming. A number are capable of secreting a calcareous tubular shell. Polychaeta are common in marine and estuarine conditions, but only a few species are truly freshwater.

2.20.2 *Oligochaetes*

Oligochaete worms (Figure 2.7(7)) range from 1 mm to several metres in length. The elongate segmented body is tapered at either end, and each segment bears several setae. These are usually S-shaped with a swelling near the middle, but they may vary greatly in form, being sharp or blunt, forked or pectinate, narrow or thick. Most oligochaetes feed on dead organic material, particularly decaying vegetation. A few feed atypically; *Chaetogaster* feeds on small invertebrates, *Ripistes* on particles filtered by its setae. Many species are freshwater, though there are many terrestrial (and some marine) forms too. Freshwater oligochaetes are mainly benthic and occur in various habitats, especially lentic ones, where they may be abundant among vegetation or in soft sediments. The Branchiobdellidae are parasitic, and attach to the exoskeleton of crayfish.

2.20.3 *Leeches*

Leeches (Hirudinea) are usually moderately large as adults, most species being 1–5 cm long. The segmented elongate body is dorsoventrally flattened, with two suckers, one at each end, the anterior one surrounding the mouth (Figure 2.7(8)). Most leeches are carnivorous and feed on other invertebrates, mainly aquatic insects, worms and molluscs, which are swallowed whole. Other leeches suck the blood of various invertebrates (e.g. insects, crustaceans and molluscs) and vertebrates (fish, amphibians, reptiles and some birds and mammals). There are about 300 species; though some are marine and a few terrestrial, the majority are confined to fresh waters. Most leeches are benthic and are active only during feeding. Bloodsucking species attach to the host only to feed, though a few species, having gained attachment, may remain there more or less permanently.

2.21 Arthropods

This is by far the largest invertebrate group, with more than one million species, and containing about 80% of all animal species. All arthropods are segmented—the segments being grouped to form a head, abdomen or other part of the body. Many of the segments bear jointed appendages, often highly modified in various ways. The entire body is enclosed in a unique exoskeleton made of chitin. This is usually waterproof and may be flexible or hardened by mineral salts to form a protective case. Members of this successful group are found in a wide variety of habitats.

2.21.1 *Insects*

This enormous class (Figure 2.8(1)), despite its lack of marine representatives, contains more living species than the total of all other groups in the animal and plant kingdoms. In all insects the body is divided into three regions—head, thorax and abdomen. The head bears distinctive sensory antennae and several pairs of mouthparts, while the thorax carries three

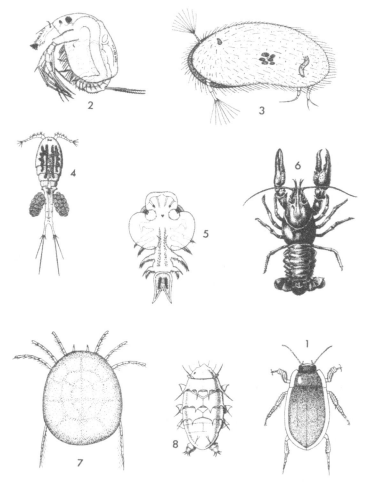

Figure 2.8 (1) Insecta (*Dytiscus*), (2) Branchiopoda (*Macrothrix*), (3) Ostracoda (*Cypris*), (4) Copepoda (*Cyclops*), (5) Branchiura (*Chonopeltis*), (6) Malacostraca (*Astacus*), (7) Arachnida (*Hydryphantes*), (8) Tardigrada (*Echiniscus*).

pairs of jointed legs and often (in adults) two pairs of wings. The abdomen
bears no walking appendages, though there may be complicated genitalia
posteriorly. In some insects, the young stages (nymphs) closely resemble
the adults (e.g. Plecoptera, Ephemeroptera) except that they have no wings
and are sexually immature. In other orders the young (larvae) may be
quite unlike the adults (e.g. Diptera, Coleoptera) and between these stages
there is another, usually dormant, phase known as the pupa. Insects are
abundant in freshwater and terrestrial habitats throughout the world, but
only a few species occur in the sea. Freshwater insects are mainly free-
living (a few are commensal or parasitic) and occur in all types of water;
there is a great species diversity in terms of morphology, behaviour and
feeding habits (carnivorous, herbivorous and omnivorous species are all
common). Most orders within the class include aquatic species, but the
following all have a major aquatic component or are entirely made up
of aquatic species: Ephemeroptera (mayflies), Odonata (dragonflies),
Plecoptera (stoneflies), Hemiptera (bugs), Trichoptera (caddisflies),
Coleoptera (beetles) and Diptera (flies).

2.21.2 *Crustaceans*

Though some groups in this class (e.g. Cirripedia, Hoplocarida) are
exclusively marine, others form a very important part of freshwater
communities.

Branchiopoda Members of this group (Figure 2.8(2)) possess flattened
leaf-like trunk limbs, made up of an exopodite and an endopodite. These
limbs usually carry numerous fine setae and may be used for respiration,
feeding and locomotion. Though some forms are over 1 cm long, most
species are only a few millimetres in length. Almost all branchiopods are
free-living and planktonic or benthic; most species are transparent and
colourless. Food consists mainly of bacteria and phytoplankton or
benthic debris. A few species are carnivorous and have modified anterior
appendages for grasping and chewing. There are four orders: Anostraca
which have no carapace, Notostraca whose carapace forms an anterior
shield over the body, Conchostraca in which the body is enclosed in a
bivalve carapace, and Cladocera where the carapace encloses most of the
trunk and limbs, but not the head. Cladocerans are extremely important;
though a few species are marine, the majority are freshwater and occur in
a wide variety of habitats throughout the world. The three other orders
are found almost exclusively in fresh water, but are largely confined to
small temporary ponds and ditches.

Ostracoda   Ostracods all possess a bivalve carapace (Figure 2.8(3)), usually strengthened by calcium carbonate, which is hinged dorsally and can completely enclose the body. Strong adductor muscles close the two halves of the carapace, which is often sculptured or beset externally with setae. Most species are only a few millimetres long and pigmented in various ways. They are mainly benthic filter feeders, but planktonic ostracods do occur, and some forms are predators, commensals or parasites. They are widely distributed in the sea and most freshwater habitats. There are about 200 living species in four orders—three of these are entirely marine, but the Podocopa includes many freshwater species too.

Copepoda   This is a large and important crustacean group (Figure 2.8(4)); most are only a few millimetres in length, though some parasitic forms may be longer. The body is segmented and composed of head (with a median eye), thorax and abdomen. The paired antennules are long, and the mouthparts modified in various ways. The biramous thoracic limbs are used for swimming, while the last segment of the abdomen bears two caudal rami. The eggs, held externally in one or two sacs, hatch into typical crustacean nauplius larvae. Copepods may be herbivorous, carnivorous or parasitic. Most planktonic and some benthic forms filter-feed on algae, while others are predaceous on small crustaceans or insects. Most parasitic crustaceans occur within this sub-class: of its seven orders, four are entirely parasitic and two partly so; the remaining order is mostly free-living. Common hosts are crustaceans, other invertebrates and fish; many of the parasites are highly modified. The copepods include some 4500 species (more than 1000 of which are parasitic), most of which are marine, but there are some extremely important freshwater genera, notably in the Calanoida, Cyclopoida and Harpacticoida.

Branchiura   Branchiurans (Figure 2.8(5)) are similar in structure to the copepods but are characterised by a great dorsoventral flattening of the body (part of which forms a shield-like carapace convering the head and thorax). A pair of sessile compound eyes and a pair of ventral suckers are also typical. All members of the sub-class (which contains about 80 species) are ectoparasitic on the skin and fins, or in the gill cavity, of various marine and freshwater fish.

Malacostraca   The largest of the crustacean sub-classes, this includes a wide variety of forms, ranging in size from a few millimetres to 30 cm or more. The eyes are stalked, and the mouthparts and antennae well developed. All segments of the trunk (which may be enclosed in a carapace) bear appendages; these may be modified for respiration, walking, swimming, feeding, egg-bearing or defence (Figure 2.8(6)). The Malacostraca includes many marine, freshwater and some terrestrial forms; of the thirteen major orders, only six occur in fresh water and these are all included in the following three superorders:

(a) Syncarida, which have no carapace but possess characteristic thoracic and abdominal appendages. There are two orders—the Anaspidacea, endemic to a few waters in Australia, and the Bathynellacea, interstitial species occurring in subterranean habitats in many parts of the world. Syncarids feed mainly on organic detritus filtered from the bottom by the mouthparts and thoracic appendages.

(b) Peracarida, which have a distinctive ventral brood chamber formed from the coxae of the thoracic limbs. Four orders include freshwater species—the Thermosbaenacea, the shrimp-like Mysidacea, the dorsoventrally flattened Isopoda and the laterally flattened Amphipoda. The simpler Peracarida are filter-feeders, but some orders feed directly on large food particles and are often scavengers.

(c) Eucarida include most of the largest crustaceans. The carapace is large and fused with the thorax; there is no brood chamber, development often including a larval stage known as the zoea. Of the two eucaridan orders, only the Decapoda (the largest crustacean order) has freshwater members, notably species of crayfish, shrimps and crabs. These are all benthic in habit and are mainly scavengers or predators on other invertebrates and fish.

2.21.3 *Arachnids*

In this large class (Figure 2.8(7)) the body is divided into a prosoma and an abdomen. The prosoma is protected dorsally by a solid carapace and ventrally by one or more plates. The appendages include a pair of chelicerae, a pair of pedipalps and four pairs of legs. Almost all arachnids are carnivorous, prey being seized, held by the mouthparts and partly digested by midgut enzymes passed out from the gut. The partly digested food is then swallowed. The ten orders in the class are almost entirely terrestrial, apart from groups within three which have become secondarily aquatic. In the Araneae (spiders) a few species occur in fresh water, where they construct shelters containing a bubble of air. In the Acari, the

Hydracarina (water mites) are found in salt and in fresh water and may be an important part of some communities. Many water mites are brightly coloured and active swimmers, preying on small invertebrates.

2.21.4 *Tardigrades*

The Tardigrada, or water bears, are all small, less than 1 mm in length. The body is short, cylindrical and blunt at either end. It is enclosed in a cuticle which may be divided into plates or drawn out into bristles (Figure 2.8(8)). There are four pairs of legs each terminating in four claws. Most tardigrades feed on plants, whose cells are pierced by sharp mouthparts and the contents sucked into the pharynx; some may be carnivorous. A few species are marine, but most are freshwater. They occur on the bottom or among plants in lakes and rivers, but the majority appear in the water films surrounding the leaves of terrestrial mosses and lichens. These forms can withstand extreme desiccation and low temperatures when they contract and shrivel, thereby passing into a state known as anabiosis.

2.22   Chordates

The Chordata is a large and successful group containing more than 40 000 living species. All are bilaterally symmetrical and characterised by the possession at some stage of a notochord—a long rod-like structure which supports the body and is usually replaced in adults by a vertebral column. Typical also of the group are gill slits and a long hollow nerve-chord, dorsal to the gut. The anterior end of this is modified to form a brain, and most chordates have elaborate sense organs and complex behaviour patterns. Members of the group have been extremely successful in marine, freshwater and terrestrial environments in all parts of the world.

2.22.1 *Lampreys*

Cyclostomes (Figure 2.9(1)) are distinguished from other chordates by the absence of jaws. The body is long, narrow and enclosed in a smooth and scaleless skin. There are no paired fins or limbs, but the tail bears a median fin, developed anteriorly as a dorsal fin. The mouth is usually sucker-like, and the head bears a single dorsal nasal opening and a pair of eyes; behind these there are seven pairs of gill openings. The poorly developed

Figure 2.9 (1) Cyclostomata (*Petromyzon*), (2) Elasmobranchii (*Carcharodon*), (3) Actinopterygii (*Coregonus*), (4) Choanichthyes (*Protopterus*), (5) Amphibia (*Triturus*), (6) Reptilia (*Alligator*), (7) Aves (*Aythya*), (8) Mammalia (*Ornithornynchus*).

endoskeleton is mainly cartilaginous. Of the two living orders, one (the Myxinoidea: hagfishes) occurs only in the sea, but the Petromyzontia (lampreys) occur both in the sea and in fresh water. Lampreys have two distinct life history stages: the larvae, known as ammocoetes, live buried in silt in fresh waters where they feed on particulate organic material brought into the mouth by a current created by cilia; most adult lampreys, on the other hand, have a well-developed sucker mouth with sharp teeth, and feed on fish in both fresh and salt water. Lampreys occur in suitable waters in the temperate areas of both hemispheres, but are absent from the tropics. As adults they can be important predators of fish, being known to have virtually eliminated large populations of certain teleosts in some areas (e.g. the Great Lakes of North America).

2.22.2 Sharks and rays

The fish belonging to this class (Figure 2.9(2)), which includes the sharks, skates and rays, form a distinct group, none of which has an air bladder, an operculum or bone in the skeleton—which is almost entirely cartilaginous. The tail is heterocercal, and the body is covered with small placoid scales—specialised in the mouth to form rows of teeth. All elasmobranchs are carnivorous (usually on fish or benthic invertebrates) and occur mainly in the sea. A few species occur in fresh water, e.g. *Pristis*, which is found in large rivers in China, India and Central America.

2.22.3 Bony fish

This group (Figure 2.9(3)), which includes the majority of successful living fishes, form the class Actinopterygii and possess not only a well-developed skeleton of bone, but also an outer covering in the skin of overlapping bony scales. Most possess an air bladder, used primarily as a hydrostatic organ and giving great efficiency and mobility in water. This is a very adaptable group and its members have specialised in many ways, developing a wide range of feeding mechanisms and occurring in most types of aquatic habitat. They are the dominant organisms in many aquatic communities. Their well-developed sense organs (including eyes, ears and chemoreceptors) and complex behaviour patterns are undoubtedly of significance in this dominance. There are at present only four living orders. Three of these (Palaeoniscoidei, Chondrostei and Holostei) contain only a few primitive forms restricted largely to fresh water in a few parts of the world. The remaining order (Teleostei) includes over 30 000 species occurring in all parts of the world as a major component of the fauna in most marine and freshwater habitats.

2.22.4 Lungfish

Living representatives of this group (Figure 2.9(4)) are the freshwater lungfishes (Dipnoi) and the recently discovered coelacanth *Latimeria*. The lungfishes are elongate with limb-like or filamentous fins and thick bony scales covering the body. There are no proper vertebrae, and the large air bladder can be used as a lung to breathe air. These fish feed mainly on small invertebrates and vegetable detritus. The three living genera (*Neoceratodus* in Australia, *Lepidosiren* in South America and *Protopterus* in Africa) are found only in certain tropical rivers which become very low or stagnant during dry weather. Under such conditions they burrow into the mud and aestivate for long periods.

2.22.5 *Amphibians*

Amphibians (Figure 2.9(5)) have much in common with the remaining chordates (reptiles, birds and mammals), notably the possession of paired lungs, two pairs of jointed limbs, usually with digits, and a well-developed central nervous system. Amphibians are characterised by the virtual absence of scales in the skin, which is well developed, normally moist and used for respiration; it often contains mucus and poison glands as well as an underlying layer of pigment cells which enable the animal to alter colour. Adult amphibians feed mainly on invertebrates, especially insects but also worms and molluscs. The aquatic larvae may be herbivorous or carnivorous. There are about 2000 different species which, though widespread in temperate and tropical areas of the world, cannot maintain themselves in many types of habitat—though they are well adapted for life in a few. Most forms are unable to survive for long unless near water, though the adults may be completely terrestrial. There are three living orders: Urodela (newts and salamanders), Anura (frogs and toads) and the limbless Apoda, which are mainly terrestrial. Most of the Urodela and Anura live in fresh water for at least part of their lives, usually as larvae.

2.22.6 *Reptiles*

Reptiles (Figure 2.9(6)), like amphibians, are cold-blooded, but differ in having dry hard skin which contains no glands. It is normally developed into bony plates or scales, and is often brightly coloured. The yolky eggs have a shell which provides protection against physical damage and desiccation. The class is widespread in temperate and tropical parts of the world. The Squamata (snakes and lizards) is the largest of four living orders but has few aquatic species, while the Chelonia (tortoises and turtles) are often, and the Crocodilia (crocodiles and alligators) are largely, aquatic by nature. The Chelonia includes about 100 species with marine, freshwater and terrestrial representatives; freshwater turtles and terrapins feed on aquatic vertebrates (especially fish) and invertebrates (particularly insects). Species of Crocodilia occur in fresh waters in most parts of the tropics and feed mainly on fish. They are well adapted to an amphibious mode of life and are important predators in some fresh waters.

2.22.7 Birds

Birds form a compact and well-known class with numerous distinctive features (Figure 2.9(7)). With mammals, they are unique in being warm-blooded, and this is part of the reason for their success. They are unique, too, in the possession of feathers—often brightly coloured—which, together with the evolution of the forelimbs as wings, have helped them to develop the power of flight. The brain is large and their behaviour often complex. The characteristic beak may be highly modified for specialist modes of feeding, and there are often other extreme adaptations of body form. Birds are an extremely successful group found in most parts of the world. Though few can be regarded as completely aquatic, some live on, and many more feed from, aquatic habitats—both marine and freshwater—and can be an extremely important part of certain communities. Some nineteen orders are recognised, and most of these contain aquatic or semi-aquatic species. Members of the following orders are made up largely of species associated with fresh waters: Gaviiformes (divers), Colymbiformes (grebes), Pelecaniformes (cormorants and pelicans), Ciconiiformes (storks and herons), Anseriformes (ducks, swans and geese), Gruiformes (rails and cranes), Charadriiformes (waders and gulls) and Coraciiformes (kingfishers).

2.22.8 Mammals

This class (Figure 2.9(8)), which includes *Homo sapiens*, is often considered to be the most highly developed group of animals in the world, and is dominant in many environments. While there is much to be said for this belief, especially as far as the human species is concerned, it is probably least true in freshwater systems where mammals are rarely prominent in communities. The class is a compact one characterised by the possession of mammary glands, fur and (in most cases) internal development of the young. Like birds, all mammals are warm-blooded. They are a very successful group with representatives in all parts of the world, from polar regions to the equator and in marine, freshwater and terrestrial environments. Though retaining basic mammalian characteristics, many species have become highly adapted to suit different environments. Among the most specialised forms are those which have become aquatic, notably the Cetacea (whales), Sirenia (sea cows) and Pinnipedia (seals and walruses)—most of these are marine, but each group has a few freshwater representatives. Of the twenty-five other orders within the class, the following also have important freshwater forms: Monotremata (including the duck-billed platypus), Marsupialia (e.g. the water opossum of South America), Insectivora (the otter shrew of Africa), Rodentia (beaver and capybara), Carnivora (otter) and Artiodactyla (hippopotamus).

CHAPTER THREE

STANDING WATERS:
LAKES, PONDS AND POOLS

Fresh waters have traditionally been divided into two types—standing waters and running waters—and this is a definite and acceptable start to their classification. Though occasional problems may arise (e.g. with canals and the backwaters of slow-flowing rivers) there is usually little difficulty in deciding whether a water belongs to the standing or the running water series. Rarely are the two unconnected, however, for many running waters pass directly into standing ones or have standing waters in their catchments, and most standing waters receive several running waters, and exit via a single running outflow.

Standing waters exhibit great variety, ranging from small shallow temporary pools, through ponds and lakes, to enormous waters which may be over $20\,000\,km^2$ in area and $500\,m$ in depth. Some of these large lakes are included in Table 3.1. Superimposed on the size and shape of a water basin are important regional differences, especially those related to geochemistry and climate. Depth is one of the most important characteristics of a lake, because on it depends the proportion of the lake's volume which receives solar radiation. The heat provided by this determines any thermal stratification and stability of the lake. Solar energy is also used in photosynthesis, which forms the basis for the productivity in standing waters. Because most energy is absorbed within the uppermost $3\,m$ of water, shallow waters absorb almost as much solar radiation as deep ones (Figure 3.1).

In catchments where precipitation is greater than evaporation, most standing waters have an outlet from which water eventually finds its way to the sea. Water in such basins is constantly renewed; consequently nutrient salts do not accumulate, and the water stays fresh. In catchments where evaporation is greater than precipitation (as in all areas of inland drainage), higher lakes are flushed periodically, but lower ones are not; the latter accumulate dissolved chemicals and are commonly called salt lakes. These are not within the scope of the present book.

Table 3.1 Physical details of some of the world's largest freshwater lakes.

Lake	Surface area (km²)	Maximum depth (m)	Volume (km³)
Superior	83 300	307	12 000
Victoria	68 800	79	2 700
Huron	59 510	223	4 600
Michigan	57 850	265	5 760
Tanganyika	34 000	572	18 940
Baikal	31 500	730	23 000
Erie	25 820	64	540
Winnipeg	24 530	19	3 110
Malawi	22 490	273	8 400
Ontario	18 760	225	1 720

In their individuality, lakes may be compared to oceanic islands (Murray, 1910); just as an island presents peculiarities in its rocks, soil, fauna and flora due to isolation by the ocean, so do lakes have individuality and peculiarities in physical, chemical and biological features owing to their position relative to catchment drainage and their separation from other standing waters by land. The endemic species of invertebrates and fish found in very old lakes (e.g. Lakes Malawi and Baikal) result from such isolation.

3.1 Origin

Hutchinson (1957) lists seventy-six different types of standing water basin in eleven major categories. A simpler classification is given by Murray (1910) where distinctions are based on the geological origin of basins. Three main types of basin are distinguished: rock, barrier and organic.

Rock basins have originated in several different ways:

(a) Solution lakes in depressions caused by solution of the bedrock, usually limestone, but sometimes sodium chloride, calcium sulphate or aluminium and ferric hydroxides. Most such lakes are created by the subsidence of surface rocks following solution and erosion of materials underneath.

(b) Volcano lakes, on the erupted sites of old volcanoes, have a single circular basin, often with no visible outlet. Sometimes the collapse of a large volcano forms several smaller basins; these may also develop on areas of lava flow.

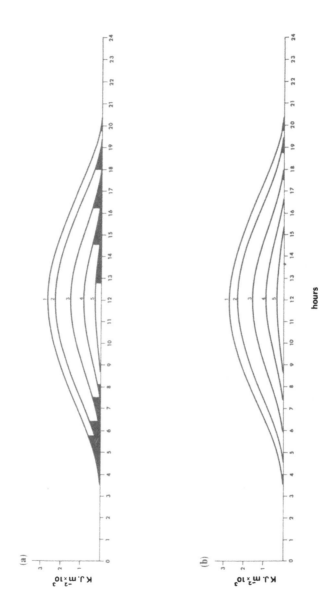

Figure 3.1 Examples of daily energy inputs from direct solar radiation for one temperate lake (Loch Lomond) (1,21 June; 2, 15 August; 3, 21 March/September; 4, 15 February; 5, 21 December). (a) Above, the northern end of the lake which is surrounded by mountains: the shaded areas indicate when there is no direct radiation because of shading. (b) Below, the southern end where there are no mountains (from Smith *et al.*, 1981).

(c) Glacial lake basins include: ice scour basins (created by ice moving over jointed rocks and scooping out depressions); corrie basins (caused by mountainside erosion due to alternate freezing and thawing at the rock face and excavation by ice on the basin floor); valley rock basins (excavated by ice action, the large ones where glaciers occupied long valleys) and drift basins (produced by ice melting on glacial drift).

(d) Tectonic lake basins are formed by movements of the earth's crust, and arise in depressions caused by faults and upward movements which alter valley drainage or cut off a previously defined marine basin. Lake basins may also form in folds on the earth's crust or by underground subsidence.

(e) Meteor basins are created by the impact and explosion of meteors, or by the compression waves around them as they approach the earth's crust.

Barrier basins are created by the damming of a pre-existing valley to form a lake; the barrier may consist of various materials. Landslides are common in mountainous areas where hard strata overlying soft ones slide into the valley—usually after heavy rain. The movement of glaciers and the deposition of materials as they pass down a main valley commonly cause a damming of streams in tributary valleys. Terminal moraines are important in creating lakes. Wind-blown sand drifted into dunes, alluvial materials deposited by streams (or by wave action in large lakes or the sea) and lava flows may all obstruct valleys to form basins.

Organic basins arise from the activities of living organisms and can be of several types. Phytogenic lakes, for instance, occur in the tundra as depressions due to differential growth of vegetation, and by damming elsewhere where massive plant growths obstruct streams. In Yugoslavia, precipitation of tufa has created several small lakes. Coral lakes arise within slightly raised and completely enclosed coral atolls, while the small dams constructed by beavers, and the larger ones created by humans, can also be considered as basins of organic origin.

Rock basin lakes of glacial, volcanic or tectonic origin start as barren waters with few nutrients and sediments, and with rocky or stony shorelines. Such lakes are unproductive and considered to be the unevolved classical oligotrophic type. In contrast, shallower lakes on glacial drift or sedimentary rocks have rich bottom sediments washed in from the surrounding land, which also supplies many dissolved salts. Such surroundings are very suitable for human habitation and for agriculture and industry. Consequently the water in such lakes becomes even richer in nutrient salts. Lakes of this type are typically eutrophic.

Once a lake basin has been created, two main forces act to destroy it. Firstly, there is erosion of the basin rim, usually at the lake outlet where there is scouring due to the action of water, sand and gravel. Secondly, most lake basins are continually being filled in by sediment washed in from tributaries, and by organic material produced within the lake. Many lakes have disappeared due to either of these phenomena or the combined effect of both. The final process of filling in is often accelerated by the invasion of macrophytes from the edge.

3.2 Physical characteristics

3.2.1 Stratification

The process of stratification, created by density differences resulting from differential heating of lake water (Figure 3.2), is of major importance in many lakes. In describing the sequence of events it is convenient to consider a lake which has been well mixed by the wind and is at a uniform temperature of more than 4°C. In the absence of strong winds and with

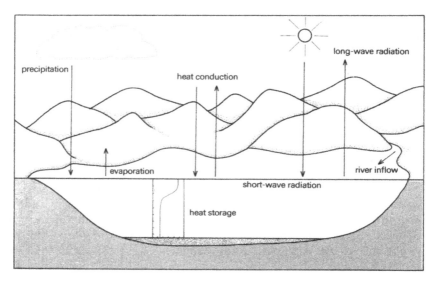

Figure 3.2   The main movement of energy in and out of a lake.

increasing solar radiation, there is a gradual rise in the temperature of the surface waters which, therefore, become less dense than the deeper layers. As this surface layer warms, an increasing density difference appears between it and the deep layer, and they become separated by a narrower layer of water exhibiting sharp temperature and density gradients. The upper layer is known as the epilimnion, the lower layer as the hypolimnion and the region of gradient between them as the thermocline (Figure 3.3). As air temperature and solar radiation decrease, the surface waters cool down again and, when temperatures in the epilimnion and the hypolimnion are similar, their waters start to re-mix.

Partly because of the anomalous expansion of water below 4°C, lakes can stratify in different ways according to the temperature regime imposed through local geography and exposure. In some lakes stratification is permanent, in some seasonal, in others intermittent, and in yet others completely absent. In almost all types, the geographical position of the lake (particularly its latitude, altitude and distance from the sea) is of

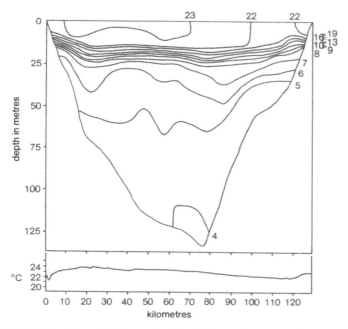

Figure 3.3 Distribution of isotherms in Lake Michigan during stratification (after Carr *et al.*, 1973).

prime importance, but superimposed on this are the combined influences of lake depth and mixing action of the wind. Standing waters may be divided into five major classes according to their stratification (Hutchinson, 1957). However, it should be noted that many lakes can be intermediate in type and may vary from year to year.

(a) Amictic (polar) lakes are permanently covered by ice, and always remain well below 4°C. They never undergo circulation. Such lakes are relatively rare, and occur only at high latitudes and altitudes.

(b) Cold monomictic (arctic) lakes never rise above 4°C in summer, when complete circulation occurs, but are ice-covered in winter, with inverse thermal stratification (i.e. water at the surface is colder than that below).

(c) Dimictic (temperate) lakes circulate completely in the spring when water temperatures rise above 4°C; they stratify during summer, and mix again in autumn, when the lake cools. These lakes are inversely stratified in winter, usually being ice-covered. This is the most complex system, and is discussed in greater detail below.

(d) Warm monomictic (tropical) lakes are thermally stratified during summer, but temperatures never fall below 4°C at other times of the year when there is complete circulation.

(e) Oligomictic (equatorial) lakes are confined to very warm areas. Water temperatures are always considerably above 4°C, and permanent stratification is normal, although this may break down intermittently due to wind.

In dimictic lakes (Figure 3.4) in spring, the water temperature is uniform from top to bottom, and wind action causes regular mixing. With increasing radiation in summer, the lake gains heat and stratification sets in, especially during calm periods. Stratification is evidenced by a gradual rise in the temperature of surface waters, compared with the temperature at greater depths. There is a rapid falling-off in the rate of temperature increase from top to bottom as summer progresses, and a definite thermocline is considered to exist when there is a difference of 1°C or more within any 1 m depth of water. At this stage there is a surface layer of water (epilimnion), about 15 m in depth and of nearly uniform high temperature. Below this is a layer of water of rapidly decreasing temperature (thermocline), and below this again a deeper layer of water (hypolimnion) of uniform low temperature.

Stratification therefore divides the lake horizontally into two parts separated by the thermocline. As the season proceeds, heat gradually

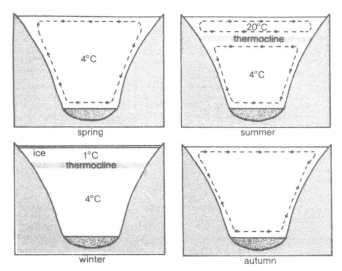

Figure 3.4 The annual thermal cycle and turnover pattern in a typical temperate lake.

transfers from epilimnion to hypolimnion, by conduction and turbulent mixing. With lowered air temperature, while the epilimnion is cooling due to loss of heat to the hypolimnion and at the surface, the hypolimnion temperature continues to rise, even though the lake as a whole is losing heat. Gradually the epilimnion increases in size, the thermocline sinks deeper and the temperature difference between epilimnion and hypolimnion decreases until the lake is again of uniform temperature from top to bottom.

During autumn the lake continues to lose heat, the temperature dropping until at 4°C the coldest water starts to float and the warmer water stays near the bottom, thus allowing inverse stratification to develop. If air temperatures remain below 0°C for long periods, the surface layers start to freeze and eventually the whole lake may be icebound. As winter passes, and with increasing solar radiation, the lake starts to gain heat again: ice melts, water is mixed by wind, and by spring the water column is once more of uniform temperature from top to bottom (Figure 3.5a).

There can also be variations in stratification due to local features of climate, topography or weather conditions (Figure 3.5b). In some lakes, even after stratification, there may be an increase in temperature with depth in parts of the hypolimnion. This is normally caused by a chemical

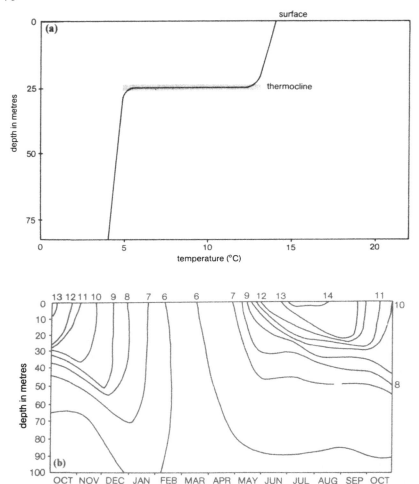

Figure 3.5 (a) Idealised depth–temperature diagram of a stratified lake. (b) Seasonal temperature data for a deep-water area of Loch Lomond during 1953–54 (after Slack, 1957).

stratification of some kind, where the decrease in density due to rise in temperature is not sufficient to overcome the increase due to dissolved salt concentration. A marked density gradient of this kind may be very stable and prevent total circulation even when the temperature of the whole water column is uniform. Such systems are called meromictic, in contrast to those with total circulation, which are holomictic. Hutchinson

(1957) refers to the region of rapidly changing chemical concentration as the chemocline.

3.2.2 *Water movement*

Some of the early studies on water movement in lakes were carried out by Wedderburn (1910) using experimental tanks. Because of difficulties in producing realistic temperature gradients in such tanks, chemical gradients with salt solutions of different densities were used. Several important discoveries emerged from these and similar experiments by other workers.

When the temperature of a lake is uniform, and a wind blows at the surface, there is a vertical circulation of the water due to currents created by this wind. A return current which provides water to replace that driven along by the wind at the surface may be detectable even at the bottom of the lake. This complete circulation is responsible for the even temperatures found in lakes between periods of stratification.

As a thermocline develops and temperature and density gradients become marked, water movement within the lake is also affected. Wind-induced return currents do not penetrate below the thermocline, but are strongest just above it. However, below this primary return current, and induced by it, there is a secondary current in the same direction (i.e. opposite to that at the surface). This secondary current is slow, sometimes hardly detectable, but nevertheless important in maintaining circulation within the hypolimnion. A secondary return current occurs near the lake bottom, even at great depths, e.g. more than 150 m in Loch Ness (Wedderburn, 1910).

The characteristic feature of water movement in rivers is that currents are mainly due to gravity, and thus constantly unidirectional. Currents in lakes may also be unidirectional, but direction depends mainly on that of the wind, and can vary in time through 360°. Constant unidirectional currents in lakes are found only where they are influenced by running waters—at river mouths, near lake outflows or in small narrow lakes which are really expansions of rivers.

In many large lakes size is virtually independent of depth, and the effect of large waves (more than 1 m in height) rarely penetrates below 20 m. In the shallower parts of lakes, however, especially near the shore where the waves are actually breaking, their movement influences the bottom. The major effect is erosion, reflected in the nature of the shoreline and littoral zones (Figure 3.6). Waves can increase the size of a lake, especially its littoral region. Erosion of the bottom brings fine material into suspension,

Figure 3.6 A typical wave-washed lake shoreline, showing extensive stony shore and fringing forest (Photo: P.S. Maitland.)

which, though it may be carried to and fro, is eventually carried into deeper water, where it settles out on to the bottom as currents diminish. The finest particles are carried furthest. In most deep lakes affected by wave action there is a gradient in sediment particle size, with stones, coarse gravels and sands at the edge, sands and some organic sediments further out and fine clays and organic silts only in the deepest water (Figures 3.7, 3.10).

As well as the obvious surface waves in lakes, there are stationary oscillations of the whole lake known as seiches. On calm days in some lakes careful measurements show that the water surface at any one point is continually and rhythmically rising and falling. Simultaneous measurements at different parts of the lake show that all the water participates in this movement (Figure 3.8). These oscillations may be simply harmonic or more complicated and, although there may be synchronisation between movements in different places, these may not be the same everywhere (Chrystal, 1910).

The range of a seiche is the vertical distance between high and low water; half this distance is called the amplitude. The simplest seiche to consider is one occurring in a long narrow lake. Somewhere near the mid-line of the lake, at a point called the node, there may be no vertical

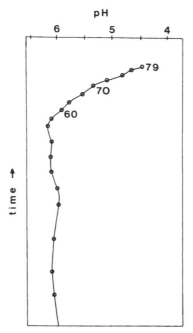

Figure 3.7 Recent changes in the pH of Lake Gardsjon, an acidifying lake in Sweden, as shown by an analysis of siliceous algae in the bottom sediments (after Andersson and Olsson, 1985). The years 1960, 1970 and 1979 are indicated.

movement at all. However, vertical movements increase in range towards one end of the lake, and simultaneous movements in the opposite direction increase in range towards the other end.

Several factors can initiate seiches in lakes. Wind blowing for some time in one direction causes surface waters to move to one end, and there may be an increase in water level at this end. This denivellation can be considerable in shallow lakes and, if the wind producing it suddenly stops, its release may give rise to a seiche. A sudden flood to one part of a lake can set up a seiche. Sudden variation in barometric pressure may also initiate one, for example by the disappearance of a prolonged depression which has caused an alteration in the level of one part of the lake. Periodic fluctuations in barometric pressure are also responsible for many seiches, and are most effective when their period is similar to that of the seiche in question.

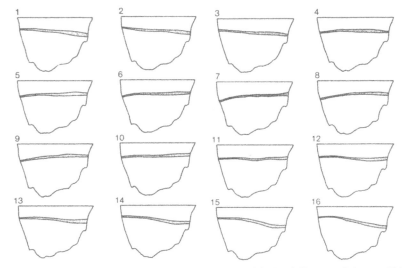

Figure 3.8   The successive position of isotherms 9–11 °C in Loch Earn on 9 August 1911, on successive hours from 0100–1700 h (after Wedderburn, 1912).

3.2.3   *Suspended solids*

Material found in suspension, in solution, or in the sedimented deposits of lakes may, regardless of its true chemical or physical character, be divided into two types according to origin. Autochthonous matter is that which has been generated within the lake itself, mainly by the growth of algae and macrophytes (Figure 3.9), the decaying parts of which add material to the lake. Allochthonous matter originates outside the lake and is subsequently brought into it, either by inflowing tributaries or by wind. Most organic allochthonous material is derived from peat, fallen leaves and other types of decaying vegetation. Pollution by humans may also be important.

Suspended material in natural waters is often referred to as seston. This is composed of both inorganic and organic material (tripton) as well as living organisms (mainly phytoplankton and zooplankton). The total amount of seston present is important in connection with the optical properties of lake water, especially the quantity and quality of radiant energy which can penetrate to different depths. Where large quantities of non-living suspended matter are present, they may have an inhibiting effect on primary production. Much of the colour of lake water is due to the nature of suspended materials.

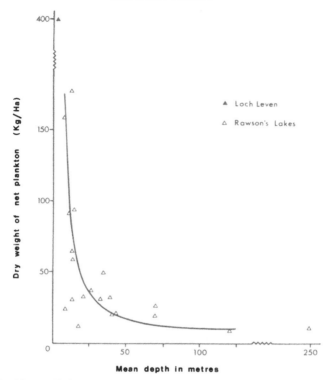

Figure 3.9   The predictive relationship between the mean depth of a lake (d) and its phytoplankton crop (p), demonstrated by Rawson (1955) for North American lakes. $p = (3765.0/[d^{1.5337}]) + 8.0$. Loch Leven in Scotland with a very shallow mean depth should predictably have a very high algal crop and this is indeed the case (from Maitland, 1979).

During complete circulation, especially in shallow waters, wind action can stir up large amounts of bottom sediments. In many lakes, large quantities of sediments are kept in suspension for long periods; in others they are stirred up and sedimented again from time to time. Material previously settled out in one part of a lake may be transferred to another quieter area (Figure 3.10), and in this way silt is continually being transferred from shallow to deep water. During stratification there may be notable differences in the suspended matter above and below the thermocline. In the epilimnion, wind circulation and the effect of light mean that the suspended content is high due to allochthonous particles or the plankton itself. There is little material disturbed from the bottom.

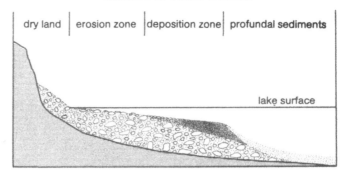

Figure 3.10   Transverse section of a typical wave-washed lake shore, showing distribution
of sediments.

This is also true of the hypolimnion, but here there is much less non-living
seston (which sediments out because of reduced water movement) or living
plankton—inadequate light for phytoplankton means little food for
zooplankton. Much of the material in the hypolimnion consists of decaying
phytoplankton and zooplankton from the epilimnion. If the density
gradient within the thermocline is great, it is not uncommon to find a
thin layer of suspended matter settled out of the epilimnion but not dense
enough to penetrate the hypolimnion.

Organisms dying in the epilimnion and other suspended matter there
are broken down by bacteria and fungi to simple dissolved materials,
which may be utilised immediately by plants, and fine particulate matter.
Larger fragments fall through the epilimnion into the hypolimnion where
they continue to decay, but at a slower rate because of reduced temperatures. Eventually, particles settle on the bottom where, if they are not
broken down within a short period, they become covered by other particles
and permanently sedimented.

In the deeper parts of lakes, such materials, originally suspended but
now settled as sediments, may initially form food and a microhabitat for
invertebrates. The activity of these animals is restricted to the upper layers
(rarely deeper than 20–30 cm), and below this the sediments are anaerobic;
here there is little activity or change either chemically or physically, other
than some compaction. Each layer of undisturbed sediment is gradually
overlain by another and yet another as the inexorable process of lake
sedimentation proceeds. Each undisturbed layer contains within it clear
evidence of the quality of the seston (and hence the lake itself) at the

time of sedimentation. Apart from the chemical aspects of such sediments, the presence of skeletal portions of many plants (e.g. diatoms) and animals (e.g. Cladocera and Chironomidae) can render a valuable picture of the previous history of the lake. Allochthonous components of such sediments—notably pollen grains—can also reveal previous events in the lake catchment. Quaternary research along these lines has greatly elucidated the history of many standing waters (Figure 3.11) and their surroundings.

3.2.4 Light

The amount of radiant energy reaching a lake and the importance of suspended and dissolved solids in controlling how far light of different wavelengths will penetrate have been discussed above. Curves produced by plotting the percentage transmission of wavelengths of different light through a layer of water 1 m deep are useful in defining the optical properties of individual waters. When comparisons of transmission curves from different lakes are made, the greatest contrasts occur among the short wavelengths, and transmission values from these are useful in categorising waters.

The depths to which rooted macrophytes and attached algae can grow on suitable substrates is largely controlled by the spectral composition and intensity of light there. Some species are capable of growing where the light intensity may be as little as 1% of that at the surface. Normally light restricts the growth of rooted macrophytes to water less than 10 m deep—often much less than this in lakes rich in plankton or those affected by suspended matter (Figure 3.12). Spence (1967) showed that in eutrophic lakes plants occur only in shallow water; they are found deeper in poor to moderately rich brown waters, and the deepest growths are in clear calcareous waters. It appears that charophytes, and to a lesser extent bryophytes, can grow in deeper water and with less light than most angiosperms.

In lakes whose waters are well circulated, the depth of the photic zone is less important to phytoplankton as the continual mixing brings each plant into the light regularly enough to photosynthesise. When the lake stratifies, however, water in the epilimnion and hypolimnion is separated and, though conditions within the former may remain the same (until nutrients are depleted), this is not the case within the hypolimnion. Here reduced circulation causes heavier species to settle out and die. Even those buoyant enough to stay in circulation may find the amount of light below the thermocline to be inadequate for photosynthesis.

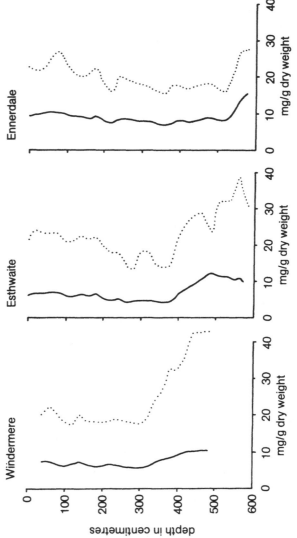

Figure 3.11 The vertical distribution of sodium (left) and potassium (right) in the deep sediments of three English lakes (after Mackereth, 1966).

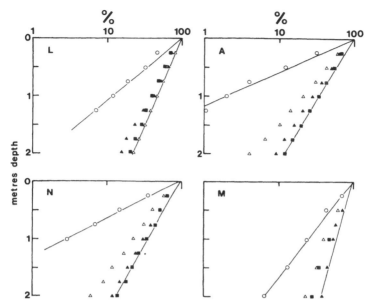

Figure 3.12 Underwater light conditions in four temperate lakes (Lochs Lomond, Awe, Ness and Morar). Penetration of blue (○), red (△) and green (▲) light is expressed as changes in percentages (%) of surface values (from Bailey–Watts and Duncan, 1981).

3.3   Chemical characteristics

3.3.1   *Dissolved gases—oxygen*

As in other aquatic systems, the quantity of dissolved oxygen is of prime importance in standing waters. Because the amounts held in water, even at saturation, are small compared with those in air, and because the rate of diffusion of oxygen in water is many times less than in air, the level of oxygen is often much more critical in an aquatic than a terrestrial habitat.

The quantities of oxygen in a lake or pond depend on the extent of contact between water and air, on the circulation of water and on the amounts produced and consumed within each system. In shallow ponds with rich plant growths but little circulation there may be violent fluctuations of oxygen with supersaturation during the day but low values at night. With adequate circulation, however, such fluctuations disappear, and in most lakes the oxygen levels remain at or just below saturation.

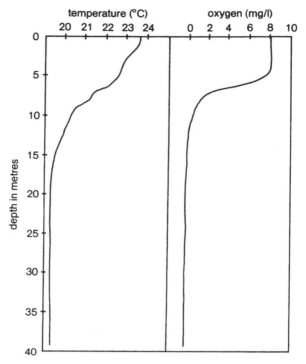

Figure 3.13 Vertical profiles for temperature and oxygen in a stratified lake—Lake Bunyonyi in Uganda (after Denny, 1972).

Thermal stratification, however, can produce a marked difference in oxygen levels, and may lead to a clear stratification of oxygen itself (Figure 3.13).

The water in well-mixed lakes is saturated with oxygen from top to bottom at a concentration which depends on temperature and atmospheric pressure. As soon as thermal stratification sets in, water in the epilimnion (which remains circulating and saturated) becomes separated from that in the hypolimnion. In the hypolimnion, though it may circulate slowly, water does not have contact with the atmosphere, and rarely are plants present to produce oxygen. Once stratification is established, therefore, oxygen in the hypolimnion starts to decrease; the rate at which this takes place depends on several factors. Temperature is important: since the oxygen is being used by fish and invertebrates whose rate of activity may depend on temperature, clearly the higher this remains during stratification the faster will the oxygen be used up. Lake productivity is also of importance: in rich lakes there are higher standing crops of organisms and greater quantities of material falling from the epilimnion than in poor lakes. Thus oxygen is used up more rapidly in rich lakes. Finally, the volume of the hypolimnion relative to the bottom area of the lake and to the epilimnion is of significance. Waters with a deep hypolimnion lose oxygen much more slowly than those with a shallow hypolimnion.

In some lakes the hypolimnion may become completely deoxygenated; H_2S and CH_4 are produced during long periods of stratification (e.g. in Lakes George and Rudolf) with fatal results for many fish and invertebrates. Differences in the loss of oxygen from the epilimnion and hypolimnion are complicated by the fact that, even though the epilimnion water remains at saturation as it becomes warmer, the amount of oxygen is reduced because of decreased solubility at higher temperatures. Thus it is not uncommon, especially in deep nutrient-poor lakes, for there to be more oxygen in the hypolimnion than in the epilimnion during early stratification and warming. Such a situation is known as orthograde, in contrast to the more normal position (clinograde) where there is a decrease in oxygen concentration with depth.

With cooling in the autumn the whole water body mixes again and becomes saturated with oxygen. If no freezing takes place, this situation continues until the next spring. If, however, the water freezes and ice cover is complete, the situation under the ice is similar to that in the hypolimnion, and the same factors affect the loss of oxygen. Thus deep nutrient-poor lakes lose oxygen more slowly than shallow nutrient-rich ones. During prolonged ice cover, anaerobic conditions may develop in the latter. However, where light penetration is good and there is active photosynthesis below the ice, high dissolved oxygen levels may be found.

3.3.2 Dissolved gases—carbon dioxide

In lakes which have undergone complete circulation, the pH and the carbon dioxide concentrations are uniform from surface to bottom. During daylight, however, there may be a contrast between shallow well-lit water and deeper water where there is little or no light. In the former, the algae and macrophytes reduce the amount of carbon dioxide and calcium carbonate (some of which precipitates), thereby increasing the pH, whereas in deeper water (where there is no photosynthetic activity) there is a tendency for an increase in the carbon dioxide and calcium carbonate

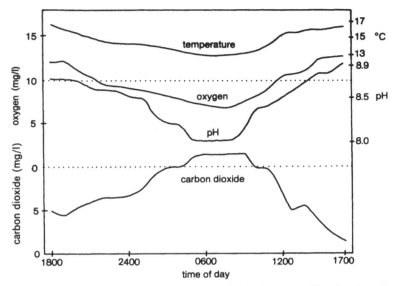

Figure 3.14 Diurnal fluctuations in temperature, dissolved oxygen, pH and carbon dioxide in the surface waters of Buckeye Lake in Ohio, U.S.A., in summer (after Tressler *et al.*, 1940).

and a reduction in the pH. A distinct stratification along these lines, however, rarely develops because of the constant turbulent circulation occurring throughout the whole lake. In most natural waters, pH and carbon dioxide in solution are more or less dependent on each other.

Following on stratification, though the situation in the epilimnion remains essentially the same, in the hypolimnion the carbon dioxide content may increase and the pH decrease. In lakes with reasonable amounts of calcium, equilibrium is quickly set up and, though the bicarbonate content increases steadily, the amount of free carbon dioxide does not increase, nor does the pH. In lakes where there is little calcium available, the concentration of carbon dioxide can increase considerably, thus lowering the pH (Figure 3.14).

3.3.3 *Dissolved solids*

As in running waters, the quality and quantity of dissolved solids in a standing water body depend on the geological nature of the drainage basin and on the influence of humans. In ponds and lakes the concentrations of various dissolved solids may vary greatly in both time and space.

Table 3.2 Generalised characters of eutrophic, oligotrophic and dystrophic lakes.

Character	Eutrophic	Oligotrophic	Dystrophic
Basin shape	Broad and shallow	Narrow and deep	Small and shallow
Lake substrate	Fine organic silt	Stones and inorganic silt	Peaty silt
Lake shoreline	Weedy	Stony	Stony or peaty
Water transparency	Low	High	Low
Water colour	Yellow or green	Green or blue	Brown
Dissolved solids	High, much N and Ca	Low, poor in N	Low, poor in Ca
Suspended solids	High	Low	Low
Oxygen	High at surface, low under ice or thermocline	High	High
Phytoplankton	Few species, high numbers	Many species, low numbers	Few species, low numbers
Macrophytes	Many species, abundant in shallow water	Few species, some in deep water	Few species, some abundant in shallow water
Zooplankton	Few species, high numbers	Many species, low numbers	Few species, low numbers
Zoobenthos	Few species, high numbers	Many species, low numbers	Few species, low numbers
Fish	Many species	Few species	Very few species, often none

In standing waters the rates of organic productivity are often closely associated with the availability of certain dissolved solids; the cycling of many of these solids is linked with the alternating periods of stratification and complete circulation which occur in lakes. The extent of deoxygenation in the hypolimnion, which affects the release of some of these salts, is itself related to the organic productivity of the lake concerned (Table 3.2). There can be a wide distinction in the biology of lakes which, however similar in their physical features, differ widely in their chemical features.

In considering the dissolved solids present in standing waters, two major types of lake can be distinguished (as discussed above): those which receive water from inflowing streams and also discharge water via an outflow of some kind, and those which only receive water but have no outflow; these represent the terminus of the catchment which they drain. The latter type of lake is essentially a saline system, not a freshwater one. Water in the former type of lake will obviously closely resemble that in its inflowing streams, and much of the value to the lake of the dissolved solids in these waters will depend on the rate at which water passes through the system.

The retention period can range from only a day or so in many small lakes on large river systems to many years in some of the larger lakes in the world. In general, the average composition of the water in lakes tends to be very similar to that in adjacent or connected rivers. Some substances are, however, accumulated within the lake by deposition or biological assimilation.

Because the dissolved salt content of standing waters is dependent on the quality of materials in its catchment area, a greater variety and much higher total concentration of ions are generally found in ponds and lakes in lowland cultivated areas lying on soft calcareous rocks than in standing waters on harder rocks in highland areas. For this reason the former are normally nutrient-rich lakes with relatively high levels of calcium, magnesium and bicarbonate, while the latter have low levels of these salts and tend to be acid in nature. Some areas are exceptions to this: waters in the Dolomites are highland but highly calcareous. There are also poor lowland lakes in some places.

In the epilimnion of lakes, high levels of oxygen and biological activity—especially photosynthesis—occur. There is some loss via solid organic material (mainly dead plankton) sinking down through the thermocline: in this way many of the dissolved solids fixed by the phytoplankton may be bound up in algae or lost through dead plankton. Major nutrients like nitrogen, phosphorus, iron, silicon and others may be depleted and so limit production or alter the composition of the algal community.

In the hypolimnion, oxygen concentration drops at a rate dependent on the productivity of the water body and other factors discussed above. In some waters, especially deep oligotrophic lakes, the oxygen level drops slowly and the chemistry of the water alters little. In others, however, oxygen disappears rapidly, reaching such low levels that several changes take place in the dissolved solids present. The bacterial community may alter to become dominated by anaerobic instead of aerobic species. The upper layers of sediment, formerly aerobic, become anaerobic and release various dissolved salts into the water. In particular, ferric phosphate precipitated out during aerobic conditions is reduced to the soluble ferrous state and liberated into the water.

In permanently stratified lakes there may be large differences in the chemistry of water above and below the thermocline. In a few lakes there is so much dissolved matter in the deeper water that even with uniform temperature chemical stratification exists, with the upper and lower layers separated by a chemocline (Figure 3.15). Thermal stratification may also exist under such conditions, the thermocline not necessarily coinciding with the chemocline. Lakes with a permanent chemocline normally have

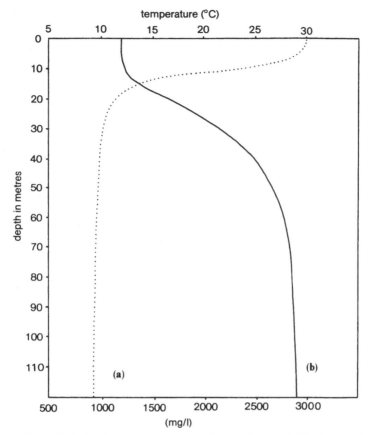

Figure 3.15 Vertical distribution of (a) temperature and (b) chloride in an idealised chemically stratified soda lake.

some source of saline water or a high rate of bacterial mineralisation near the bed leading to a high dissolved solid content in the water there.

3.4 Biological characteristics

3.4.1 *Habitats*

Standing waters are extremely variable in character, ranging from temporary puddles through shallow bog or marl ponds to enormous open lakes which may be very shallow (e.g. Lake George) or extremely deep (e.g. Lake Tanganyika). Within a single standing water, several habitats may be represented, especially in shallow waters, where conditions can range from sheltered areas with a muddy substrate and abundant plant growth to exposed shores with a clean substrate of bare rock or stones and no plants. Even in very small water bodies there may be great differences in the conditions in the middle of the pool compared with the edge.

The biological characteristics of open waters may be classified into two broad categories—pelagic and benthic systems—the latter being subdivided into littoral and profundal types. The species composition of communities of all these types is greatly influenced by the nutrient status of the water concerned. The water surface itself forms a unique habitat for specialised organisms as discussed below.

The pelagic habitat is that of the free water away from the influence of shore or bottom substrates, while the benthic habitat is associated with the substrate of the lake, extending from the shoreline into deepest water. The littoral habitat may be defined as extending from the shoreline out to deeper water until wave action and the light needed for benthic plant growth are insignificant. The benthic profundal region is considered to occupy the rest of the basin beyond this.

The pelagic habitat proper occurs only in larger standing waters where there is sufficient area and depth for the water to be uninfluenced by the substrate. There are two main zones within the pelagic region—one near the surface, the other in deeper water. In stratified lakes there may be a definite boundary between these in the form of the thermocline. Conditions prevailing in the upper pelagic region (in some cases the equivalent of the epilimnion) include high light intensities (ensuring adequate photosynthesis), regular contact with air (maintaining stability of dissolved gases), and, in exposed lakes, wave action near the surface. In the lower pelagic region, light is poor or absent, there is no contact with air and there are no violent waves or currents—though there may be mixing due to return currents. In stratified lakes there is normally little movement below the thermocline, and the dissolved gas content there may vary considerably during the year.

In the benthic littoral region, the light intensity is adequate for rooted macrophytes and attached algae, and there are many similarities to the upper pelagic as far as circulation is concerned. The extent of water movement determines the type of substrate and vegetation here; in exposed areas this may consist of bare rock, boulders, stones or clean sand with little or no vegetation, while in sheltered situations the bottom may be of mud with rich growths of macrophytes and attached algae. In profundal benthic areas waves have little effect and the substrate consists of fine silts or clays. The only motion at the mud surface is due to return currents. Chemical conditions vary greatly according to stratification and other factors, and can range from a situation like that at the lake surface to complete deoxygenation.

3.4.2 Microhabitats

A wide variety of microhabitats occurs in standing waters; the nature and importance of these is becoming more evident as detailed studies appear but the definition and classification of the various microhabitats is still a matter of debate, and only major concepts and types within standing waters are discussed here.

The habitats described above contain within them a variable number of microhabitats: the open water of lakes, for instance, contains essentially only one, whereas the benthic littoral region may include many. One interesting habitat found in all fresh waters is the water surface itself, and the organisms linked with this are known as neuston. Those associated with the upper surface of the water film are known as epineustic, those with the lower surface of the film as hyponeustic. These organisms rely on the surface tension of the water film itself to support them, most of them showing modifications for this purpose. Many water skaters (Hemiptera) have elongate legs ending in pads which support them on the surface film, over which they can run swiftly. The chrysophycean *Chromatophyton rosanofii* supports itself on the water surface in a gelatinous capsule. Among higher aquatic plants many floating forms (e.g. *Lemna* and *Salvinia*) are cosmopolitan: all have foliage which is water-repellent and has a high proportion of air, so that it floats easily. All such plants absorb their nutrients direct from the water. Among hyponeustic forms, the rhizopod *Arcella* and the cladoceran *Scapholeberis* are able to glide under the surface film, feeding on particulate material adhering to it. Some macrophytes (e.g. *Utricularia*) float just under the water surface where their position is maintained by buoyancy, not attachment to the surface film.

Within the pelagic zone there is little in the way of subdivision into microhabitats. Many of the organisms have buoyancy or swimming devices to maintain themselves in the water; plankton proper are subject to movement by the water mass while nekton are larger animals capable of active independent movement. All such organisms can exist in the epilimnion indefinitely, but only a number of them for limited periods in the hypolimnion, especially when this is anaerobic. Many pelagic organisms exhibit the phenomenon of vertical migration.

It is among the benthic areas of standing waters that the greatest variation, and consequently the greatest number of microhabitats, is found. Substrates range from bare rock through boulders, stones, gravels and sands to fine clays and organic muds. Among these occur benthic algae and macrophytes. These add further complexity to the niches. Some substrates, such as smooth bedrock, are simple; others (such as bare sand) are relatively so, offering the sand surface itself, its aerobic upper layers and the anaerobic deposits beneath. Bare mud offers an essentially similar number of niches, and is the major substrate in the profundal zone. The most complex situations are found among mixed areas of stones, gravels and macrophytes; these offer a variety of niches above and below stones or leaves, among gravel or silt, and within the stems or leaves of the plants themselves. It is in dealing with complex situations such as these that any arbitrary system of microhabitat classification finds itself in difficulties.

3.4.3 *Communities*

The communities of standing waters relate to those of the principal habitats and include neuston, plankton, nekton, profundal benthos and littoral benthos. As in small running waters, the question of permanence is important in small shallow standing waters, especially in warm areas. In such situations the flora is reduced (many of the macrophytes are semi-aquatic forms) while the animals are often good burrowers or can encyst for long periods. Insects which are only partly aquatic may make up an important part of such communities.

3.5 Classification

Various schemes for the classification of standing waters have been suggested and, though even the best of these is open to criticism, most of the systems so far presented have been more successful than similar schemes for running waters (see Chapter 4). This is mainly because each body of standing water is an entity with characteristic and reasonably uniform physicochemical and biological components. Running waters, on the other hand, can rarely be precisely defined and often exhibit a wide range of conditions and communities within a single system. The classification schemes proposed for standing waters so far have been based on a variety of parameters including type of origin, physical (especially thermal), chemical and biological characteristics.

Classification according to origin has been discussed, and three main types of basin distinguished: rock, barrier and organic. Within each of these are found further subdivisions. Though this classification is a useful one and extremely practical in that most lakes fall distinctly into one class or another, it has two main disadvantages. Firstly, many lakes have changed dramatically since they were first formed, and are still doing so. This difficulty is solved to some extent in the classification suggested by Pearsall (1921) by arranging lakes in an evolutionary sequence. Secondly, lakes with entirely different origins can be similar from an ecological point of view and vice versa.

Lakes can be arranged according to their superficial areas, the volumes of water they contain, their mean or sometimes maximum depths, their latitude, altitude or salinity. All these classifications must be regarded as more or less artificial and, while useful in comparing a large number of lakes, are not of great value from the ecological point of view, mainly because the lakes form a complete series in each case. Any divisions, therefore, tend to be arbitrary ones.

Whipple (1898) classified lakes into three types according to their surface temperatures.

(a) Polar, where the surface temperature is never above that of maximum density (4°C).
(b) Temperate, where the surface temperature is sometimes above and sometimes below that of maximum density.
(c) Tropical, where the surface temperature is never below that of maximum density.

Whipple also included in this system three further subdivisons according to bottom temperatures.

(a) Where the bottom temperature is rather constant, near the point of maximum density.
(b) Where the bottom temperature is less constant and undergoes annual fluctuations.
(c) Where the bottom temperature is rarely very different from that at the surface.

Forel (1904) also divided lakes into three types based on their bottom temperatures.

(a) Polar, where the temperature of the deep water varies from and below that of maximum density.

(b) Temperate, where the temperature of the deep water varies above and below that of maximum density.

(c) Tropical, where the temperature of the deep water varies from and above that of maximum density.

Each of these three types is subdivided into two categories—shallow and deep. Shallow lakes are defined as those which have a variable bottom temperature, and deep lakes as those with a constant bottom temperature.

Another classification according to thermal characteristics has already been discussed, i.e. the system developed by Hutchinson and Loffler (1956) where there are five major classes of water: amictic, cold monomictic, dimictic, warm monomictic and oligomictic. This is probably the most satisfactory of the thermal systems.

Because of the importance of temperatures in lakes, especially as far as stratification is concerned, systems of classification involving thermal characteristics are obviously potentially useful ones. In addition, the subdivisions are real ones with only a few intermediate cases, and this makes such systems even more valuable. The disadvantage of classifications relying solely on thermal characteristics is that no account is taken of the nutrient content and consequent productivity of such waters, and it is quite possible for two lakes which are very similar thermally to differ chemically, and consequently to be dissimilar ecologically.

Several workers have attempted to relate the distribution and abundance of plants and animals to one or more chemicals, and to classify waters in terms of their chemistry. Such systems are often useful but usually rather arbitrary. Total hardness or alkalinity is frequently used for classifying into nutrient types. Spence (1967) defined waters which have up to 15 mg/l of calcium carbonate as nutrient-poor, those with 15–60 mg/l as moderately rich, and those with more than 60 mg/l as rich.

One of the most useful biological classifications of standing waters was originally suggested by Thienemann (1925) and later elaborated by others. This scheme suggests three major types of open water—oligotrophic, eutrophic and dystrophic—and some of the major differences between these are listed in Table 3.2. Oligotrophic lakes are nutrient-poor, usually deep, clear lakes which never have oxygen deficiency; the chironomid midge *Tanytarsus* is often dominant and the culicid midge *Chaoborus* absent. Eutrophic lakes are nutrient-rich, usually shallow, turbid lakes which may have an oxygen deficiency in deeper water at some times of the year. The chironomid *Chironomus* is normally dominant and *Chaoborus* is present. Dystrophic lakes have variable amounts of nutrients but high amounts of humus, making the water brown; they are usually shallow or only moderately deep, and may show oxygen deficiencies in deeper water. *Chironomus* and *Chaoborus* may be present, but in low numbers. Mesotrophic lakes are intermediate in character between oligotrophic and eutrophic ones. Though this classification is extremely general and in some ways arbitrary, it has proved its value over a long period, and the types described are widely used.

No system of classification is ideal. Nevertheless many are of value, and if the systems themselves are defined and understood correctly, they provide useful methods of categorising and comparing lakes. It is probable that some combination of the systems discussed above is the most acceptable way of defining a lake quickly, for example some lakes can be edaphically oligotrophic but others only morphologically so. To describe two lakes as eutrophic may be insufficient when one is deepish, dimictic with a heavy precipitation of marl, while the other is shallow and oligomictic. In spite of complications, however, the development of schemes of classification is much more likely to be successful for standing than for running waters and the development of more sophisticated schemes is likely to be useful.

CHAPTER FOUR
RUNNING WATERS:
RIVERS, STREAMS AND TRICKLES

Running and standing waters may be differentiated as follows:

(a) A consistently unidirectional current is found in all running waters, but not to any degree in standing waters.

(b) Stratification rarely occurs in running waters (due to the current), but is a characteristic feature of many standing waters.

(c) In running waters, physical and chemical conditions change gradually from source to mouth, and the difference in many factors may be great between these. Conditions in standing waters are normally much more homogeneous.

(d) Running waters are normally shallow and have long, often complex, narrow channels. Standing waters may reach great depths, but mostly have simple broad basins.

(e) Constant erosion is characteristic of running waters, and materials so removed may be transported considerable distances, often right out of the catchment concerned. Erosion does occur in standing waters, but is rarely severe; eroded materials usually remain within the same basin.

(f) As a further consequence of erosion and deposition, most running waters increase the length of their channels with age, as cutting back to the source and meandering on the flood plain proceed; in standing waters, materials are constantly being deposited, tending to fill in the basin and eventually obliterate it completely.

(g) Currents in running waters are normally stronger than those in standing waters.

There are many kinds of running waters, sometimes several of them occurring, interconnected, with a single drainage system. This presents difficulties in classifying running waters into absolute categories. The range covered within the series includes small trickles and seepages (often

Table 4.1 Physical details of some of the world's largest rivers.

River	Discharge (100 cumecs)	Length (km)	Catchment (1000 km^2)
Amazon	1724	6274	6133
Zaire	396	4667	4015
Yangtze	218	5794	1943
Mississippi	176	6260	3222
Yenisey	174	4506	2590
Lena	155	4281	2424
Parana	149	4500	2305
Ob	125	5150	2484
Amur	96	4667	1844
Nile	28	6695	2979

temporary in nature), ditches, larger fast-flowing streams and rivers, large slow-flowing rivers and canals (Table 4.1).

The geology of the catchment underlying a running water has a strong influence on physical and chemical characteristics. Further, the nature of the bedrock and soils affects the character of the aquatic substrate and also relates to the rate of erosion, and hence succession, of that running water. The suspended solids present may also be determined by the degree of erosion, and they in turn affect light under water. The flow characteristics of running waters are also connected to geology, notably in the control exerted by rock and soil formations, and the relationship between ground water and surface waters. The flow pattern of running waters depends largely on the nature of this relationship. In addition to controlling the quantity of ground water, local geology also exerts a strong effect on its quality, notably in connection with the dissolved solids present.

Most water on earth is in circulation, within what is known as the hydrologic cycle (Figure 4.1). The energy utilised within this cycle comes mainly from the sun. Water evaporates from both land and sea to be reprecipitated, usually somewhere else. On most parts of the land, precipitation exceeds evaporation, and run-off towards the sea occurs.

4.1 Physical characteristics

4.1.1 *Current*

The velocity of currents in running waters depends on the nature of their gradients and substrates. In contrast to standing waters, wind has little

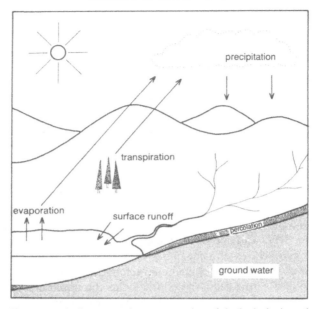

Figure 4.1   A diagrammatic representation of the hydrologic cycle.

influence on currents in running waters, except occasionally in large
slow-flowing rivers. Current velocities in running waters are at their
maximum in waterfalls and their minimum in pools; great extremes can
occur within a single system, occasionally close together, though the fastest
currents are normally found in the upper reaches and the slowest in the
lower reaches.

Information on gradients, and thus on the current velocities, in a running
water may be obtained by examining its profile. In most running waters
the gradient is steep in the upper reaches for some distance below the
source; in the middle reaches the gradient becomes much less severe, while
in the lower reaches it tends to be very slight, especially near the mouth
(Figure 4.2). Such a profile alone may give valuable information, not only
on present conditions within the system, but also on the stage of
development which the water has reached in relation to its succession
(Figure 4.3).

Current velocity within any stretch of a running water depends also on
the shape of the channel and has a regular pattern in a transverse section
(Figure 4.4). Near the water surface, current velocity is reduced by surface

Figure 4.2 The slow-flowing reaches of a meandering lowland river, showing zonation of emergent and floating leaved vegetation (Photo: P.S. Maitland).

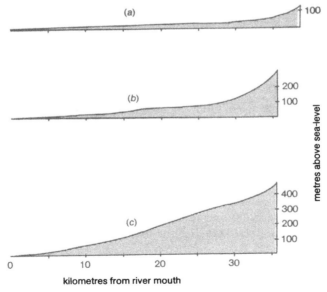

Figure 4.3 Contrasting running water profiles: (a) River Arun, England, (b) River Tyne, Scotland, (c) River Ebbw, Wales.

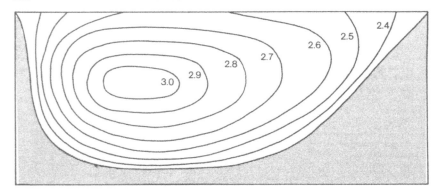

Figure 4.4   A transverse section of a running water channel, showing contours of equal current velocity. The figures represent metres per second.

tension, while near the substrate it is also reduced due to friction; in a typical channel, the maximum velocity normally occurs at a depth of about 0.3 of the total depth from the surface. In experimental water courses it has been shown that:

(a)   The velocity at 0.6 of the total depth is usually within 5% of the mean velocity.

(b)   The mean velocity is usually from 0.85 to 0.95 of the surface velocity.

(c)   The average of the velocities at 0.2 and 0.8 of the depth is usually within 2% of the mean velocity.

Uniform channels occur rarely in nature, and so variations from this pattern are common. Local water velocities are affected greatly by the shape and size of the channel, especially in relation to the degree of meander. Ice cover greatly increases friction at the surface and the maximum velocity in such a situation is found at about the mid-depth. Currents near the bottom are variable and dependent largely on the nature of the substrate. Flow here may vary from a smooth and laminar one, where the substrate is flat and regular, to a highly turbulent one, where large stones and boulders make up the bed.

4.1.2   *Suspended solids*

The erosion, transportation and deposition of solid materials within a running water are closely linked to current velocity. The erosion of a water

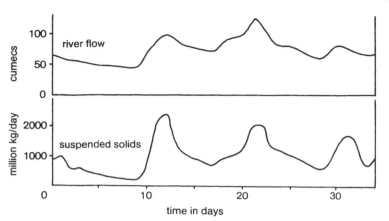

Figure 4.5 The discharge of suspended solids and water in the Missouri River (U.S.A.) at Waverley (after Meinzer, 1942).

course depends largely on current velocity and the character of the channel bed (Figure 4.5). Materials removed from the bed by the current are mainly carried along in suspension, but heavier particles may actually be rolled along the bottom, thereby causing further erosion by scouring (Figure 4.6). The deposition of materials is also dependent on the current, heavier particles gradually dropping out as velocity decreases.

The suspension of small particles—usually fine silt and detritus—in water causes turbidity, a factor of great importance but which may vary from place to place. In highland streams, which have rocky beds, turbidity is usually low, while in silted lowland rivers it may be high. Though in some running waters turbidity may be constant, in others it may vary widely according to season or spates. The suspended solids causing turbidity are important in altering light penetration; they may also serve as food for various invertebrates. Where the current is slowing, many suspended solids may settle on the bottom, affecting the substrate there (Table 4.2).

4.1.3 Light

As in other ecosystems, light is a major factor in the ecology of running waters in connection with plant photosynthesis and growth. The penetration and intensity of light in any water are closely related to turbidity there (Figure 4.7). In addition to the scattering by suspended

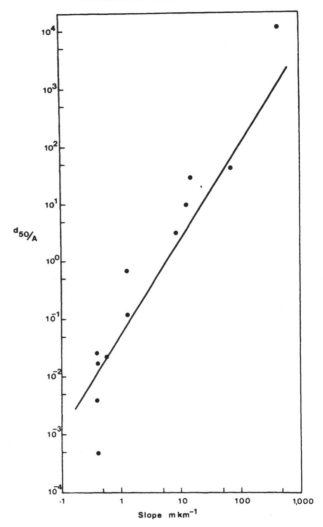

Figure 4.6 The relation between channel slope and the ratio of median particle size to catchment area in a single river (the River Tay). The dots represent sampling sites along the river, the solid line indicates the relation shown by Hack (1957) for rivers in North America (from Maitland and Smith, 1987).

Table 4.2 The diameter of objects moved and the types of substrate created by different current velocities (after Nielsen, 1950).

Current velocity (cm/s)	Diameter moved (mm)	Substrate type
10	0.2	Mud
25	1.3	Sand
50	5	Gravel
75	11	Coarse gravel
100	20	Pebbles
150	45	Small stones
200	80	Stones
3000	180	Small boulders

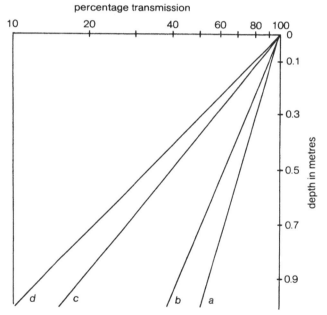

Figure 4.7 Transmission of light through different river waters in England: (a) River Test, (b) River Colne, clear, (c) River Colne, turbid, (d) sewage effluent (after Westlake, 1966).

solids there is also loss due to absorption by the water itself and substances dissolved in it. The productivity of a running water can depend therefore on its turbidity. If the water is clear (i.e. of low turbidity) or relatively shallow, adequate light to permit photosynthesis can reach the substrate. If the water is very deep or turbid, primary production at the substrate may be prevented by inadequate light. In extreme cases, turbidity may reduce the light by as much as 90% in the first 2 cm of water depth.

Compared with the information available on the penetration of light in various standing waters, there is a dearth of equivalent data for running waters. Yet light is just as important here. Light entering a standing water is normally the major source of energy there, via the plant photosynthesis, whether by phytoplankton or attached algae or macrophytes on the bottom. In a running water, however, unless adequate light for photosynthesis reaches the substrate for plants there to utilise it, it may be lost to the system concerned, certainly in that stretch and perhaps altogether if there is little phytoplankton present.

4.1.4 *Temperature*

Due to the absence of stratification in running waters there is normally uniformity of physical and chemical characteristics in any one stretch; clearly this is true of temperature. Another major difference between standing and running waters is that the temperature of the latter follows that of the air much more closely than that of the former. In certain cases, where the temperature of the running water differs greatly from that of the substrate over which it flows, the temperature of the water just above the substrate may differ from that near the surface.

Most running waters originate in high land and flow with varying velocities down towards an inland lake or the sea. The temperature of such waters is lowest at the source, gradually becoming higher downstream. Many integrated factors affect river temperatures, however, among which the most important are:

(a) *Origin.* Running waters have two main sources: firstly, surface water, the temperature of which is close to that of air and varies accordingly; secondly, ground water, which has a constant low temperature. The relative proportions from these two sources will influence the temperature characteristics of a running water.

(b) *Depth.* Water temperature is related to relative exposure to air; deeper waters are less affected by air temperatures than shallower waters.

(c) *Substrate.* The shallower a water, the greater will be the effect of the substrate on turbulence near the surface. The greater this surface turbulence, the greater will be the effect of the air temperature on that of the water.

(d) *Tributaries.* The temperature of water entering a river from a tributary has an influence related to the relative sizes of tributary and river at their junction.

(e) *Exposure.* Solar radiation is a major source of heat for running waters, and so well-exposed stretches have higher temperatures than those subjected to shade.

(f) *Time of day.* Imposed on the seasonal temperature regime is a circadian cycle caused by the diurnal variation in solar radiation and the varying temperature of the air between day and night.

4.1.5 *Run-off*

Run-off from land is the excess of precipitation over evaporation there. It is usually related to a catchment and refers basically to water travelling on or beneath the ground surface into a channel. Water precipitated on land may undergo various processes:

(a) It may evaporate and move to reprecipitate elsewhere.

(b) It may move as surface water into a running water and so pass to the sea.

(c) It may be absorbed by the roots of plants and transpired by their leaves into the atmosphere again. Local vegetation is extremely important in connection with run-off; those plants with shallow roots depend mainly on current rainfall, but deep-rooted plants rely on, and may affect, sources of ground water. The rate of soil erosion into a river channel is also linked to the quality and quantity of local plant cover.

(d) It may travel directly as surface run-off into the sea.

(e) It may percolate into the soil and remain there as ground water. Some of this will be evaporated near the surface or transpired by plants, but most will eventually find its way into a nearby running water.

The contribution of surface and ground waters to the flow in drainage systems varies according to a number of factors—especially local geology and climate. Running waters fed mainly by surface water have a variable flow and may spate with each heavy rainfall. Those fed largely by ground

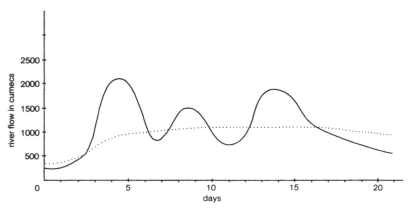

Figure 4.8 Contrasting patterns of flow in a single river system—the Stillwater River,
U.S.A.,—above (solid line) and below (dotted line) a reservoir.

water are usually regular in flow, though they may spate occasionally
after prolonged rain, when large amounts of surface run-off occur
(Figure 4.8). Run-off may also be modified by freezing, where water is
stored as snow or ice which gradually melts and helps to sustain stream
flow.

Water run-off is a primary agent in shaping the earth's surface; those
of secondary importance are wind, frost, ice and vegetation. Surface run-off
moving in thin sheets is laminar in flow, and its velocity is directly
proportional to gradient. As water depth increases and depressions occur,
flow becomes turbulent and its velocity is proportional to the square root
of the gradient; thus as flow increases, the ground surface is more and
more eroded and gullies may be formed. The vegetation present has a
major effect on the degree of erosion taking place, the nature of the root
complex being especially important. Steep unprotected slopes of fine soils
are those which erode most quickly and give rapid surface run-off into
water courses, with highly variable flows there. Level land, on the other
hand, with good plant cover and granular soil, permits infiltration and
ground water recharge, thereby producing well-sustained stream flows.

In most cases, run-off lowers land level, though in certain low-lying
places it may raise it where eroded materials are deposited. Each running
water tends to reduce its basin to a level from which there will be no further
change. As in standing waters, therefore, there is the phenomenon of
ecological sucession, with a regular sequence of change from one
environment to another. Due to erosion, each running water constantly

cuts back to its head waters; as the gradient becomes less steep in the lower valley, the river begins to meander, thereby eroding its banks and widening its valley. In this way, conditions characteristic of the lower parts of a system tend to move towards its source. This movement is accompanied by a migration of associated lentic plant and animal communities. The speed of such migrations is dependent on rates of erosion and is usually a slow process.

4.2 Chemical characteristics

4.2.1 *Dissolved gases*

Of the dissolved gases present in running waters, oxygen is the most abundant and important. In most waters the concentration of oxygen is high, due to turbulence and mixing, and in normal situations it is near saturation. Though in standing waters great variations in oxygen content may be found, in running waters there is seldom present more than 100% of the saturation value of dissolved oxygen. Low concentrations usually indicate organic pollution. In some thickly vegetated slow-flowing waters, however, there may be a marked circadian fluctuation in oxygen; this increases after sunrise to midday and then decreases again. The variation is due to plants' photosynthesis during the day, which counteracts all respiration. During the night both plants and animals continue to respire, however, and no oxygen is produced. In extreme situations, the variation in oxygen content may be considerable, e.g. 36–169% saturation in one river (Butcher *et al.*, 1927). In general, the amount of oxygen present is related to the current, the water temperature, and the presence of respiring plants and animals.

Other gases of importance to aquatic organisms (e.g. carbon dioxide and various decomposition gases) tend to be scarce in running waters due to constant turbulence of water and its frequent contact with air. Such gases are not discussed here, though some dissolved solids found in ground water are important, and discussed below.

4.2.2 *Dissolved solids*

The concentrations of dissolved solids in running water may vary in time and space. Apart from interference by humans, the quality and amounts of dissolved solids depend on the geology of the drainage basin. Running and standing waters are similar in this regard, and each may influence

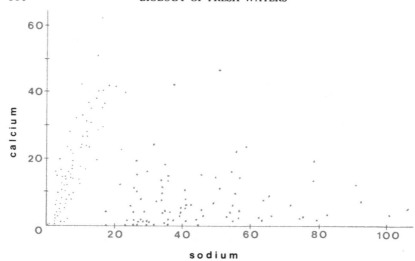

Figure 4.9   Plots of calcium and sodium content (values in mg/l) of waters in an oceanic island region (Shetland, ×) and in an inland region (Tayside, •) of Scotland.

the other in the same catchment. The dissolved solids present in a river may vary greatly from source to mouth, usually increasing in a downstream direction; the effect of rainfall on their concentration is marked, especially where ground water is important. Increased flows tend to dilute, and decreased flows to concentrate, the amount of dissolved solids in a running water.

Along with the 19 000 km³ of water which runs into the sea from the land each year are carried over 1 000 000 000 kg of dissolved solids. Even before reaching the ground, rain water contains appreciable amounts of certain dissolved solids, usually to the extent of about 30–40 mg/l (Figure 4.9). After falling on the earth, water begins to react with materials in the ground. The quality and quantity of solids dissolved from the ground depend on the character of the soil and rocks themselves, the length of time they are in contact with the water, and the nature of substances already in solution. Dissolved carbon dioxide in the water often assists in dissolving certain materials. Water from igneous rocks is usually low in dissolved salts, containing only small amounts of calcium, magnesium, sodium and potassium as bicarbonates, sulphates and chlorides. Silicon is normally the commonest constituent of such waters. Water from sedimentary rocks, on the other hand, contains higher amounts of dissolved solids, notably calcium bicarbonate but also magnesium and sodium in the form of bicarbonates or chlorides.

Ground waters often differ markedly from surface waters in dissolved gases and solids. Due to the lower oxygen content of soil, ground waters normally have less oxygen than surface waters at the same temperature; in contrast, the carbon dioxide content of ground water is usually higher than that of surface waters. Associated with the high carbon dioxide content in limestone regions, the concentration of calcium bicarbonate in the ground water may also be very high. Once such water has emerged from the soil, however, the carbon dioxide rapidly decreases, and excess lime may precipitate; in some places these deposits are large. Similarly, ground water which has a low oxygen but a high carbon dioxide content may contain large amounts of ferrous bicarbonate; when such water emerges and makes contact with air, ferric hydroxide is precipitated. Such deposits account for the well-known rust-coloured sediments associated with some trickles and seepages. Iron bacteria are often associated with such deposits.

4.3 Biological characteristics

4.3.1 *Habitats*

Running waters are variable in character, and a range can be found from swift cascading waters to slow-flowing reaches where conditions are similar to those of standing waters. This variation is also found in the organisms, some of which are characteristic of lotic habitats, others of lentic ones. Those exclusive to and most typical of running waters are found only in lotic waters. Among such organisms are some algae, and many stoneflies (Plecoptera), blackflies (Simuliidae) and caddisflies (e.g. Hydropsychidae).

In most running waters, lotic habitats tend to occur in the upper reaches and lentic ones in the lower reaches (Figure 4.10). The degree of divergence between the two habitats depends mainly on the current, which itself depends on the gradient and surface of the channel bed. Lotic and semi-lentic habitats may occur alternately in a running water as riffles and pools; the latter are rarely truly lentic, for they may be scoured out periodically during high spates. Backwaters (usually old channels) connected to running water systems are, on the other hand, normally completely lentic in character.

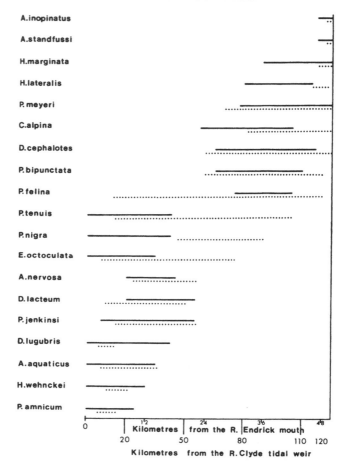

Figure 4.10 The distribution of a number of invertebrate species of limited occurrence in two different rivers, the River Endrick (—) and the River Clyde (···) (from Maitland, 1979).

Though similarities exist between a rapid-flowing stony river and a wave-washed lake shore, there are differences, however, as far as some of the organisms are concerned. The main feature is that the wind-induced currents in the littoral area of lakes have a backward-and-forward motion, and the same water remains essentially in one place, while in rivers the current is moving continually in one direction and so the habitat is constantly bathed by new water. This normally brings with it a continuous supply of fresh materials, notably nutrient salts and food particles.

One of the problems confronting organisms in lotic habitats is that of maintaining position against the current. Due to the variation found in running waters from source to mouth, it is important that each species is able to maintain itself within certain reaches and avoid being carried out of its tolerance limits; many algae and invertebrates have evolved mechanisms for attaching themselves to the substrate so that they will not be swept away. They are then able to use the current to advantage, and many invertebrates have evolved complex feeding mechanisms for filtering food particles from the water. Less obvious, but equally successful, methods of resisting strong currents are to avoid them by hiding under stones or by building heavy cases—as do some caddis larvae (Trichoptera). Other adaptations such as streamlining, flattening and rheotaxy are considered below. In many stream insects the behaviour of the adults is so adapted that the females fly upstream before ovipositing, thereby counteracting the tendency of the aquatic stages to drift downstream.

In habitats where the current is strong and the bottom unstable, erosion occurs; here, many of the organisms may be killed or swept away. Few are adapted to withstand the powerful forces during extreme spates. As the current lessens downstream, silt is deposited and smothers all firm substrates if it builds up over a period of time. Such deposits are, however, unstable in any floods which may occur and organisms in them must be adapted to such variable conditions. Algae tend to be much less important in such habitats than in those further upstream where the water is shallow and the bed stony, but strong growths of macrophytes may occur; these root firmly in the silt, further deposits of which accumulate around their roots. Most invertebrates found in these conditions burrow into the silt to escape unstable conditions at the surface. Fish are normally independent of the bottom and able to swim against the current or find shelter from it among weed beds, where too a variety of microflora and invertebrates occur.

Due to the number of parameters involved, it is difficult to give a satisfactory classification of the habitats which are found in running waters and the exact characteristics of each. More research is required before this can be attempted adequately; it is possible, nevertheless, to give some idea of the main biotopes found in running waters and their general characteristics without defining these too exactly. It is normally feasible to place any running water habitat into some category, and a general

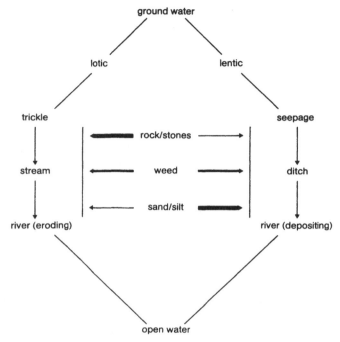

Figure 4.11   A schematic representation of the relationships among running waters and their substrates.

outline of a scheme is shown in Figure 4.11. The basic division is into the two main habitat types: lotic, where the current velocity is fast, and lentic, where it is slow. Within each of these categories a number of types is found, each depending primarily on the size of the water concerned. With lotic conditions there is a range from trickles through streams to fast-flowing rivers; with lentic habitats, the range covers seepages through ditches to slow-flowing rivers.

Within each category three basic substrates can be found:

(a) Rocks and stones, most common where the current is fast.
(b) Weed, whose occurrence depends on the species, but is most abundant in lentic situations.
(c) Sand and silt, usually associated with lentic conditions.

Most running waters start off as, or have strong connections with, ground water, itself a specialised habitat, while larger waters have open water in which may be found plankton. Habitats in the running water series which have been created by humans include drainage ditches and canals: these are essentially similar in character to natural ditches and slow-flowing rivers respectively.

4.3.2 Microhabitats

The importance of microhabitats has been realised only recently. As with the description and analysis of basic habitats themselves, much work has yet to be done before adequate diagnoses of the variety and characteristics of microhabitats in running waters can be made, and only a brief account of some of them is given here.

As in standing waters, many running waters contain a variety of micro conditions, while others include a few, occasionally only one, true microhabitat. Smooth bedrock in a stream consists essentially of a single uniform microhabitat; as soon as a piece of weed attaches itself or a crack appears in this bedrock, other conditions are produced and one or more further microhabitats are available. One of the commonest types of habitat in running water is a substrate of stones of different sizes lying among gravel; such an area has within it several microhabitats, each potentially occupied by different organisms. A single stone in the current presents several surfaces to organisms (Figure 4.12). Anteriorly, the current is moderate and slightly turbulent—this surface is fully exposed to molar agents and is less suitable for many organisms than the top or sides of the stone, where the current, though faster, is laminar. Some organisms which can withstand the current find conditions suitable here; for example,

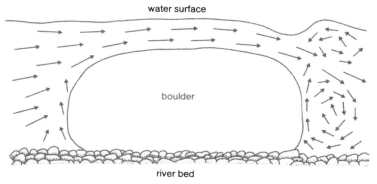

Figure 4.12 The distribution of currents around a boulder in running water.

plants are exposed to the light and least liable to silting, while certain animals (e.g. the blackfly *Simulium*) are able to filter food particles from the water. At the back of the stone the current velocity is less and, though it is turbulent here, organisms unable to tolerate the faster flows may find conditions suitable. Underneath the stone, further niches are available. If the current can flow right under the stone, the small channels through which it passes form ideal places for net-spinning caddis larvae (Trichoptera) to establish. If, on the other hand, there is no current, silting is inevitable and organisms associated with silt appear. As with bedrock, any crevices or plants which appear on the rock may offer further sets of conditions. The gravel over which such stones may lie affords a further microhabitat, often in contact with yet another—ground water.

Among beds of vegetation several microhabitats occur, the nature of which depends on the species of plant involved and the size of the clump. Microhabitats found within a clump of a typical running-water macrophyte such as *Ranunculus fluitans*, for example, include the broad surfaces of the larger leaves and stems, the narrower surfaces of other leaves and stems (a variety of current conditions would be found among both these, normally faster nearer the outside of the clump and slower on the inside), the insides of the leaves and stems (often occupied by boring animals), the basal portion of the plants where detritus and silt have collected, and the root system growing within silt on the river bed.

Bare sand or silt presents a relatively simple type of habitat, and other than the distinction between the surface of the substrate and the deposit itself there are no clear-cut microhabitats. Nevertheless, a gradual transition of conditions of oxygen, carbon dioxide and organic matter occurs from the surface of the substrate downwards, and the organisms inhabiting the substrate may show definite differences in their vertical distribution.

4.3.3 *Communities*

The types of community found in running water are inevitably associated with the habitats there, and only four basic types of community are discussed here: ground water, fast-flowing water (lotic), slow-flowing water (lentic), and free water (planktonic). In smaller running waters permanence is relevant and creates a further subdivision of community type. Typically, the flora in running waters which are temporary in nature is reduced and contains a high proportion of semi-aquatic forms. Among the animals, characteristic features include the ability to burrow (and thus escape desiccation), to encyst, or to be terrestrial for part of the life history. For this reason, insects are often dominant in such communities.

(a) *The ground-water community.* The habitat which this occupies ranges from minute interstitial crevices among sands and gravels to large underwater caves. The main environmental features are the absence of light and the stable temperatures. True plants are absent, and animals are dependent for energy on plant material carried by currents from the earth above, or each other. The ground-water fauna is unique in many ways, and due to the constant and rather unusual environmental conditions many primitive types of animal occur. The majority are almost completely blind and most forms are small (e.g. nematodes, hydracarines and crustaceans) though a few larger animals do occur (e.g. fish and amphibians).

(b) *The lotic community.* Since current velocity is the dominant factor in lotic habitats, the biology of most organisms there is geared to it in some way. Higher plants rarely occur abundantly under extreme lotic conditions, though certain mosses (e.g. *Fontinalis*) will tolerate high current velocities. Algae growing in swift waters have specialised hold-fast cells for adhering to the substrate (e.g. *Lemanea* and *Batrachospermum*). Most lotic fish are excellent swimmers; invertebrates are less efficient in this respect and can rarely swim well against currents—they tend therefore to be bottom dwellers, depending on the substrate for anchorage or shelter. For this reason many are forced to be sedentary for long periods—a disadvantage overcome by the development of filtering devices. Such animals (e.g. simuliid blackflies and hydropsychid caddis larvae) are among the most characteristic of the lotic community.

(c) *The lentic community.* In the absence of strong currents the deposition of sediment becomes a major factor in the lentic habitats. Macrophytes may play a dominant role, and slow-flowing waters may support luxuriant growths which occasionally choke the channel; macrophytes also afford shelter and food for many invertebrates and fish. Decaying organic matter in the sediments supports populations of protozoans, rotifers, nematodes and various larger burrowing invertebrates (e.g. oligochaete worms, bivalve molluscs and chironomid midge larvae). Most fish are mobile, and even among the lentic invertebrates some good swimmers occur. Figure 4.10 illustrates the degree of difference which can be found between lotic and lentic communities in a single river.

(d) *The planktonic community.* True plankton is absent in lotic conditions but may develop in the lentic lower reaches of large rivers such as the Volga and Mississippi; common groups of organisms found in such plankton are chlorophyceans, diatoms, protozoans, rotifers and various small crustaceans. In long rivers, slow-flowing water may take many weeks to reach the sea, and plankton has ample time to reproduce and increase in number. In his classic study of river plankton, Kofoid (1908) found that the organisms present were derived from a number of sources, that the plankton was subject to extreme fluctuations in quality, and that, though it had few species peculiar to it, it was characterised by the large numbers of benthic forms present. Eventually all plankton in running waters dies when it is carried into the sea. Lakes in the catchments above running waters produce plankton which is carried out at their outfalls; the numbers of these rapidly decrease along the running water below, providing food for a rich and characteristic population of invertebrates there.

Drift materials in running waters include not only true plankton and dead particulate matter, but also living benthic organisms carried along by the current. A wide variety of organisms is found in normal drift, and most members of the benthos occur at one time or another. The presence or absence of many benthic animals in the water is related to their activity in the substrate and many show definite circadian rhythms in their tendency to drift (Figure 4.13).

4.4 Classification

Attempts to classify running waters have been of two kinds: firstly, the division of the system into definable zones (from source to mouth) and, secondly, the separation of river systems into definable types. Each type of classification presents its own problems.

Various schemes of zonation for running waters have been devised; most relate to just one or a few factors—for example, a single group of plants or animals. Harrison and Elsworth (1958) based their initial scheme of zonation on physical features such as substrate, while Tansley (1939)

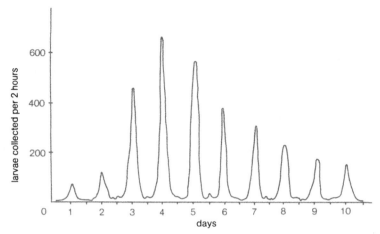

Figure 4.13 The numbers of mayfly (*Baetis*) larvae found in regular samples of drift from a small stream in Sweden (after Muller, 1965).

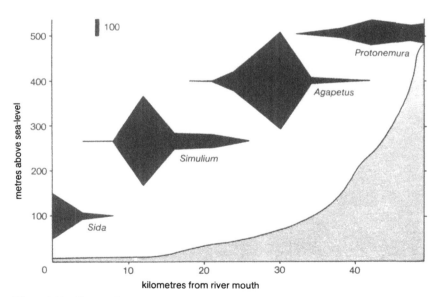

Figure 4.14 The numbers (per 10 minutes) of locally distributed invertebrates collected at twelve points along the length of the River Endrick, Scotland.

classified rivers into zones on the basis of their vegetation. Schmitz (1955) recognised various invertebrate zones in running waters, while one of the common methods of differentiating zones in rivers is the distribution of common fish species. Such schemes, originally described for European rivers by Thienemann (1912), have been adapted for rivers in the British Isles by Carpenter (1928). In general, any rigid scheme of zonation for organisms in running waters, however clear-cut it may be, is rarely valid for other groups there, or even for the same group in another water. The general theme of change in biotic associations from source to mouth in a running water is one of transition rather than zonation (Figure 4.14). The

Figure 4.15 The classical hierarchy of stream ordering. Numbers indicate the order of respective segments. The waters concerned are second- (left) and fourth- (right) order streams.

main value of characterising zones is that, in a general sense, they may be useful for descriptive purposes.

There have been various attempts also to classify whole running water systems; Ricker (1934b) classified on physical and chemical characteristics, and Carpenter (1927) on the type of origin of each system. Lagler (1949) suggested a very arbitrary system based on the average density and weight of invertebrates in a standard area. One of the most valuable discussions on the classification of running waters is that of Berg (1948), who points out that running waters cannot be classified like standing waters because they are not uniform entities, but are systems which change from source to mouth. It is not possible, therefore, to undertake ecological groupings

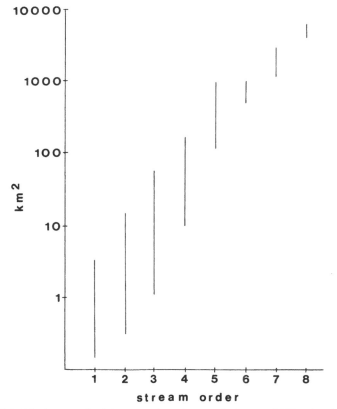

Figure 4.16 Maximum and minimum catchment areas of streams of different orders in the Tayside region of Scotland.

of running waters but merely of certain reaches within them, and the unit in such a system will not be an entire running water channel, but a stretch of one within which the environment is uniform, i.e. a habitat. The problem of classifying running waters is thus one of describing these habitats and their organisms; there is still a dearth of information about these, for the fauna and flora of few running waters have been studied in detail from source to mouth. In relation to such studies, the concept of stream order hierarchy is a useful one (Figure 4.15), with considerable validity in relation to size and other factors (Figure 4.16).

CHAPTER FIVE

FIELD STUDIES:
SAMPLING IN FRESHWATERS

Methods used in ecological investigations are clearly of importance, and the objectives of any problem being investigated must be clarified at the outset. Only then should the methods available be studied closely and appropriate ones selected. An investigator must be constantly alert for faults in methodology, and aware of the accuracy and limitations of particular techniques or equipment being used. In most ecological studies a variety of methods have to be employed to measure different parameters; each one of these has its attendant problems and resultant variations in accuracy.

The most accurate method is not necessarily the one to be selected. It may well involve the use of expensive and delicate equipment which is more difficult to use than alternative apparatus, which, though less refined, gives a sufficiently accurate result for the problem under consideration. In many cases knowledge of the accuracy of a method is more important than the attainment of a very high standard. Several other considerations are also relevant; an obvious one is whether or not it is feasible to obtain the ideal equipment and the conditions in which it will be used—important where field work in all weathers is concerned.

Consideration should also be given at the outset to the methods of recording to be employed. Often the matter is one of personal choice with regard to notebooks, card recording or other methods. However, it is important to take into account the eventual use to which results will be put. If, for instance, a particular set of records is going to be analysed by computer and the program for this is known beforehand, then much laborious effort and time can be saved by recording results according to the needs of this program from the start. The theme of methodology is efficiency, and the use of filing systems and computers is an important aspect to consider when tackling any ecological problem.

The accurate and permanent recording of field data can be difficult in poor field conditions or under water. The variety of plastic materials now available has helped to overcome many of the difficulties formerly experienced with the use of paper and pencils or pens. A useful recording system can be made up by using loose sheets of plastic (preferably the type used by draughtsmen) of a standard size, clipped to a smooth board. By writing on such sheets, one at a time, using an old ball-point pen or similar instrument, a permanent impression can be left regardless of the amount of water present. The technique will work under water when scuba diving, but is more laborious than a standard notebook and pencil and need only be used in extreme conditions.

5.1 Physical

5.1.1 *Sediments and bathymetry*

The quality of the bottom deposits found in natural waters is of fundamental importance to the ecology of animals and plants there. The

Figure 5.1 A modified Ekman grab used for sampling soft bottom sediments (Photo: Fisheries Research Board of Canada.)

equipment used for sampling sediments is often the same as that used for benthic invertebrates. Grabs and corers are indispensible, especially the latter, and a variety of efficient models is now available (Figure 5.1), some of them capable of sampling bottom sediments to a depth of several metres. Deep-coring samplers such as these are generally less useful for biological problems, however, because of the narrow diameter of the core obtained. They are ideal for obtaining material for sediment analyses—especially where investigations of the vertical distribution of deposits are concerned. Pollen and other types of analysis on the sediment samples obtained with such corers have resulted in invaluable information on past events in many standing waters.

For the sounding of underwater depths, a simple sounding line, essentially a graduated line (made of a wire which will not stretch) with a mushroom weight at one end, is extremely useful; most of the original work on the bathymetry of lakes was done by this accurate but rather laborious method (Figure 5.2a). It is still of use in many situations, often in conjunction with a sounding pole for shallow-water work. A method of estimating position within the water body is also essential, and there are now accurate electronic and optical methods for fixing position. The development of sophisticated echo sounders has meant that, for deeper waters especially, an efficient and rapid method of producing numerous depth transects is available; such sounders are now widely used in oceanographic and limnological research (Figure 5.2b).

Regular readings of water level provide useful information about the water regime of a lake or river. Water level can be measured quite simply by the establishment of a graduated post at some convenient point in a water body—preferably near the edge where it will always be touched but never submerged by the water. The position of such posts in streams is important, especially where stream gauging is concerned, the ideal situation being where there is a uniform channel with no obstructions to the flow either upstream or downstream. Posts should be levelled accurately, so that their graduations are known heights above sea level. Several sophisticated flotation methods of measuring water level have been developed; these are usually highly accurate, and equipment can be constructed to give regular readings throughout the year. Many of these meters are simply modifications of the original index limnograph (Figure 5.3a) evolved over a century ago, and are of major importance for flow recording in running waters and the study of seiches in lakes.

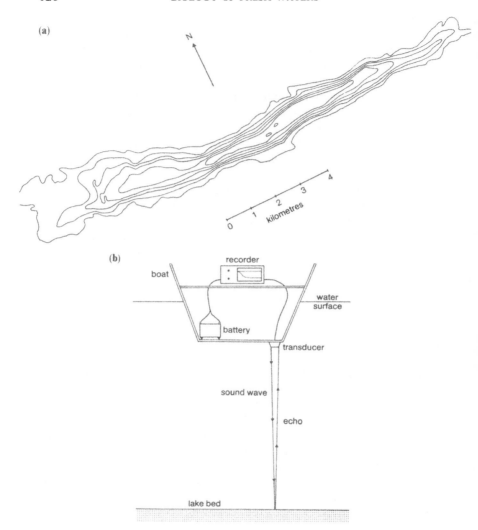

Figure 5.2   (a) Bathymetry of Loch Morar in Scotland as recorded by a hand line (after Murray and Pullar, 1910), and (b) a modern echo sounder.

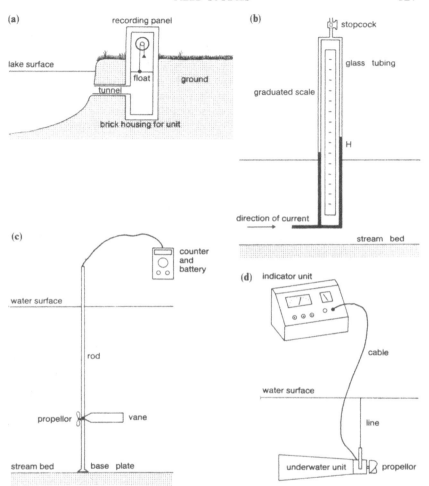

Figure 5.3 Instruments for measuring water movement: (a) index limnograph, (b) Pitot tube, (c) rod current meter, (d) direct-reading current meter.

5.1.2 *Movement*

In both running and standing waters the direction and speed of horizontal movement is critical biologically, and considerable effort has gone into developing methods of measuring such factors accurately. In simple situations float methods are useful; measurements of surface current flow can be obtained by timing a float (weighted to be almost submerged) over a known distance. Such floats can be disturbed by wind and eddies. Rod floats, which are weighted so as to float vertically in the water with only the end of the rod above, serve to give an estimation of the average velocity of the water from the surface to the bottom of the rod. In some waters submerged floats are used to determine underwater current velocities.

Originally, simple tube methods were used to measure current velocities in streams. The Pitot tube (Figure 5.3b) is basically an L-shaped tube with both ends open; when this is held vertically with the lower opening directly against the current, water enters and rises up the tube to a height h above the water level outside. The velocity of the water is calculated by using the formula $v = \sqrt{2gh}$; various improvements have been made to this simple apparatus, which can give accurate results if used properly. Another essentially similar instrument is the Bentzel velocity tube.

Several sophisticated current meters are now available which use propellors driven by the current; these meters are now the most accurate and widely used method of current measurement. In principle, they work on the basis of the propellor being held under water and the drive initiated by the current is transferred to a revolution counter, read at known intervals (Figure 5.3c). The data so obtained are applied to a formula to calculate the actual velocities. Most of these instruments are highly sensitive and must be calibrated carefully. The majority have a guide vane to keep the propellor directly into the current. In shallow water, such meters are mounted on a rod, but in deeper waters they are lowered on cables. In such situations the much older Ekman current meter, which records both current rate and direction, has been widely used (Figure 5.3d).

The development of fluorescent dyes (e.g. rhodamine B) and appropriate fluorometers to detect them has led to a very successful means of determining water movements, velocities, dilution rates and circulation patterns. Fluorometers have been increasingly used in the analysis of water movement—in both running and standing waters—and have superseded accepted methods for some investigations. Radioactive tracers have also become useful for some studies.

5.1.3 Light

Light is a factor of fundamental importance to the energy flow pattern within living communities, but its accurate measurement under water presents problems. A variety of photosensitive elements has been used in its measurement, including emission photocells, thermopiles and the

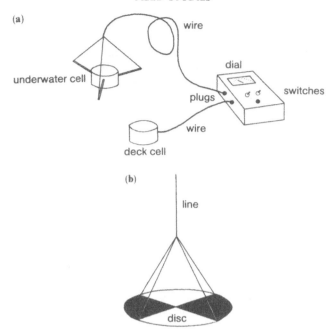

Figure 5.4 (a) A photometer, (b) Secchi disc.

selenium rectifier cell; the last-mentioned is used very commonly now since it is cheap, sensitive and easy to adapt for underwater purposes. Measurements can be recorded to depths associated with 1% of the suface intensity and with suitable amplifiers to some 0.0002%. In practice, the photometer is lowered into the water (preferably off a boom to prevent shading effects from the operator or a boat) and the photocurrent is recorded at suitable depth intervals (Figure 5.4a). Due to varying rates of light extinction in natural waters, a selection of filters is employed to isolate spectral bands, since different wavelengths are absorbed at different rates in natural waters. It is normal practice when measuring light under water to measure it also at the surface; this means that any changes in surface illumination (e.g. due to cloud cover) during the period of study can be compensated for. The combination of surface and underwater readings allows absolute intensities at various depths to be calculated.

Ways of expressing light penetration under water vary, but the most common measures referred to are the vertical extinction coefficient and the percentage transmission per metre. These may both be obtained from a semi-logarithmic plot of light intensity against depth.

A simple method of measuring the transparency of water is to use a Secchi disc (Figure 5.4b). This is a black and white disc some 20 cm in diameter lowered into the water on a line until it just ceases to be visible. This depth is measured and provides a convenient method of comparing the transparency of different waters; the technique is still widely used in limnology. The percentage transmission of light through a standard tube containing a water sample is a more accurate but involved method of comparing transparencies.

5.1.4   Temperature

Measurements of temperature are extremely important in limnology. Many field instruments are available, depending on the situation and accuracy required. It is often adequate to measure to the nearest 0.1°C, the main points to consider being the accuracy and speed of operation of the equipment in use. In shallow water it is convenient to use a mercury thermometer inside a water bottle (e.g. a Ruttner bottle) being used to collect water samples. In deeper water it may be necessary to insulate such a sample before it is brought to the surface. For general information on temperatures a simple maximum and minimum thermometer (Figure 5.5a) may be employed; even readings as seldom as once weekly can provide valuable information on local temperatures.

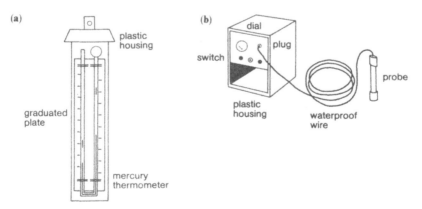

Figure 5.5 Temperature measurement: (a) maximum-and-minimum thermometer, (b) thermistor.

The reversing thermometer has been used in many classical temperature studies. In use, it is lowered to the required depth and a weight (known as a messenger) is sent down, causing a reversal of the thermometer through 180°, thus breaking the mercury column; the thermometer is read after hauling it to the surface. Its use is very time-consuming, however, and it is often advantageous to operate several thermometers at once to obtain simultaneous readings at different depths. In recent years, however, the development of the thermistor (Figure 5.5b) and its modern derivatives has overshadowed other methods, and these are now widely used for measuring temperature. Thermistors rely on a sensitive element (e.g. a sintered mixture of metallic oxides) whose electrical resistance changes rapidly with temperature. Combined with a power source, a bridge circuit and galvanometer, the resultant instrument provides an efficient and rapid method of measuring temperatures under water.

5.2 Chemistry

Many methods are available for collecting samples of water for chemical analysis from natural waters; the simplest of these is a weighted bottle lowered on a line tied both around the neck of the bottle and (about 15 cm further up) to the stopper. When the bottle has reached the required depth, a jerk on the line removes the stopper and allows the bottle to fill. The main disadvantages of this method are that it will not function in deep water and that contamination of the sample takes place as the bottle is drawn up to the surface. The latter disadvantage is overcome in the Casella sampler (Figure 5.6a) which is, however, still limited in the depth to which it can sample.

Flexible tubing is now widely used in sampling. Where an integrated sample of the water column above a particular depth is required it is sufficient to lower a large diameter (ca 2 cm) plastic tube, weighted at its free end, to the required depth. A screw clamp is now used to close the other end, and the lower end is raised by means of a line; the sample is then run into a bottle. If a large sample of water is required from a standard depth, a pump of some kind can be connected to the sampling tube. In shallow water down to about 8 m, a simple hand pump is adequate, but for greater depths something more powerful is needed; one simple method is to use an air-lift pump. Here, a narrow-bore tube runs down the inside of the sampling tube to about 1 m from its free end. Compressed air or nitrogen is then forced down this tube by connecting it to a compressed

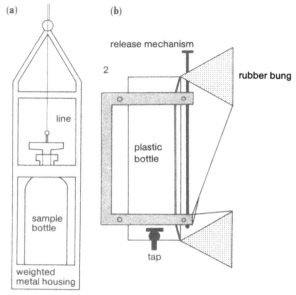

Figure 5.6 Water sampling bottles: (a) simple bottle sampler, (b) van Dorn sampler.

gas cylinder; the bubbles rising from its open end and up through the larger tube lift steady amounts of water to the surface and into a bottle.

Several useful types of sampling bottle are available for collecting water from a known depth. Among the most widely used are those of Ruttner, Friedinger and van Dorn (Figure 5.6b). These bottles consist of large wide-diameter tubes open at both ends; they are lowered to the required depth on a line. A messenger sent down this line causes both ends of the tubes to be closed in some way; the bottle is then raised to the surface and the sample removed. As well as being used for sampling water, such bottles are also of importance in taking quantitative samples of plankton.

Most chemical analyses of water are carried out in the laboratory and are not of relevance to field methods. A variety of texts describe the methods of analysis for natural waters (e.g. American Public Health Association, 1960; Taylor, 1958; Mackereth, 1963). For certain parameters, however, it is important to have field readings, and simple methods and apparatus for this have been devised (Brown and Flaton, 1943).

The recent introduction of oxygen probes and meters has enabled numerous oxygen readings to be carried out directly, and the method is used in combination with a thermistor probe; together these permit rapid measurements of temperature and oxygen profiles in lakes. Such equipment has superseded the older laborious methods. Field methods have also been developed for determining several other parameters (e.g. pH, conductivity and alkalinity). Standard meters are now available for these, and allow accurate and quick readings in the field. For most chemicals it is sufficient to bring back an uncontaminated water sample and analyse it in the laboratory. In larger chemical laboratories automatic analysing equipment is now standard and can cope with a large number of samples in a short time.

5.3 Biology

5.3.1 *Decomposition*

The role of micro-organisms in ecological processes has been recognised for some time but it is only in recent years that its importance has been defined with some precision. Much of the functioning of biological systems through decomposition, chemosynthesis and photosynthesis is dependent on the ecology of bacteria and other micro-organisms in these systems, and this in turn has a profound effect on the circulation of nutrients. However, the quantification of the role of these organisms has proved difficult (Sorokin and Kadota, 1972) both theoretically and practically and many of the methods in this field are relatively new or in the process of being developed.

The biological fixation of nitrogen is important in many aquatic systems and can be measured in various ways. One of the simplest of these is to use the reduction of acetylene to ethylene which serves as an index and can provide quantitative measurements of nitrogen in the field. One of the earliest applications of the acetylene reduction method in the field was by Stewart *et al.* (1967) who studied blue–green algae in lakes. The method is sensitive, as well as being simple and cheap and is described briefly here.

Samples are exposed in vessels which are usually small vaccine bottles or similar containers. Usually samples (of known volume) are concentrated before measurement. The larger filamentous blue–green algae can be concentrated with an ordinary phytoplankton net (70 meshes/cm) but bacteria and unicellular algae are recovered with membrane filters. The concentrated sample is thoroughly mixed and 1 ml samples are then pipetted into bottles. Air is removed from the bottle by pressure evacuation or gas replacement and the reaction is initiated by injecting C_2H_2 into the vessel. Samples are then incubated *in situ* or under standard conditions on shore. After a known time, samples are inactivated by the addition of sulphuric acid and gas samples are removed and analysed by gas chromatography. It is normal to use controls (without C_2H_2) and to replicate samples within each experiment. Periodically it is important to use ^{15}N in the experiment in order to establish a valid ratio of C_2H_2 reduced to N_2 reduced.

Bacteria, yeasts and moulds are known to play a major part in the breakdown of organic matter in aquatic environments. The decomposition processes are of two types: (a) the hydrolytic breakdown of high polymers into compounds of low molecular weight, and (b) the non-hydrolytic breakdown (mineralisation) of small organic molecules—generally accompanied by the consumption of oxygen—producing inorganic compounds used as plant nutrients. In order to measure the rate of these reactions, samples can be incubated in the field and oxygen consumption or tracers measured. The principles involved are similar to those outlined below for the measurement of photosynthesis, but for bacteria and other small micro-organisms, samples are filtered through a membrane filter before incubation.

Various other methods of measuring microbial activity are available. The mineralisation rate of organic matter in waters and sediments is best measured using radioisotopic methods. However, the uptake of organic matter (e.g. labelled glucose and acetate) can also be used to give a measure of the rate of microbial decomposition of organic matter and various methods are available to measure other aspects of aerobic and anaerobic decomposition (Stumm and Morgan, 1981).

The actual numbers of bacteria can be determined by direct microscopic counts on a membrane filter after appropriate staining. The biomass of micro-organisms in natural waters is mainly estimated by various direct and indirect techniques of determining the numbers and sizes of bacterial cells. Most of the important field techniques are described by Sorokin and Kadota (1972).

To estimate the *in situ* production rate of microbes, several field methods are used, as follows. (a) The direct measurement of the increase in the number of micro-organisms on membrane filters or in isolated samples. (b) The rate of increase in the number of micro-organisms growing on slides immersed in the water. (c) The measurement of metabolic activities of micro-organisms and the subsequent calculation of production rates. (d) The estimation of the rate of multiplication of micro-organisms in continuous culture systems.

5.3.2   *Photosynthesis*

Primary productivity in fresh waters is assessed by measuring the rate of photosynthesis over a known period of time. The first method used was a measure of the oxygen evolved, but later Nielsen (1952), using radioactive techniques, developed a method for measuring the amount of inorganic carbon incorporated.

In the oxygen method, plant material to be investigated is placed in a clear glass bottle, full of bubble-free water. This is left in light for a known time, after which the oxygen content of the water is measured. Under these conditions oxygen is evolved during photosynthesis, but respiration is also carried out by plants, animals and bacteria in the sample. To compensate for this, an identical sample in a dark bottle, from which light is excluded, is treated in the same way. The difference between the values from the dark and light bottles gives a measure of the amount of oxygen which has been evolved, and this is directly proportional to the amount of carbon incorporated.

The radioactive method also uses dark and light bottles, and is more sensitive. A known amount of radioactive carbonate or bicarbonate is added to each bottle and, after incubation, the amount incorporated in the plant cells is measured; the value from the dark bottle compensates for fixation of carbon in the absence of light.

5.3.3   *Attached algae*

Attached algae grow on various substrates, from solid rock to soft mud and the surfaces of plants. The method of sampling algae depends rather more on the nature of this substrate than on the algae themselves. Qualitative samples can be obtained from hard substrates by scraping or brushing and collecting the material; brushing is usually more satisfactory, since even crevices are cleaned out. For soft substrates, various suction devices have been developed.

Quantitative sampling involves the isolation of a known area of substrate, and the removal of algae within this into a container. Where the substrate itself can be removed from the water, and the algal community is sufficiently attached to be unaffected, then it is relatively simple to isolate a small area, say $1 \, cm^2$, by means of a cylinder or box. This should have a sharp edge which is pressed firmly against the substrate. Everything round the outside is then scraped away, the sampler removed, and the sample delineated by it scraped into a container.

Where it is not possible to avoid loss when lifting the substrate from the water or where (e.g. with solid rocks) it is impossible to do so, then sampling is done under water. Several samplers have been designed for this, e.g. on firm substrates, that of Douglas (1958). This is a tube of known cross-sectional area with a flexible leading edge which fits closely against the substrate. A stiff brush fitted inside the tube is then used to detach the algae; these are then sucked up by means of a rubber bulb, and the whole sample is thus removed from the water. Small corers (e.g. Jenkin) can be used quantitatively on muds and sands, but subsequent quantitative removal of algae from the substrate is difficult. For some quantitative estimates the sample can be dried and algae removed for weighing. Many are broken and damaged by this method, however, and it is only useful where gross estimates are required.

Studies of attached algae have been carried out in both field and laboratory by observing their growth on various artificial substrates; glass or plastic are the most popular materials for this purpose (Figure 5.7a). The technique involves placing standard slides in batches in the water and examining these regularly. Though providing results of interest, such techniques have the inherent disadvantage of not knowing how relevant experimental growth is to that on natural substrates.

5.3.4 Macrophytes

Methods of sampling terrestrial plant communities are commonly adopted in aquatic situations. For counts of individual plants or their removal for weight or other measurements, standard quadrats are useful; these vary from 0.1 to 1 m² in area. They are readily constructed in a collapsible form from angle metal, preferably aluminium since it is light, non-corrosive and easily seen under water. They can be used in shallow lakes and rivers, or in deeper water using scuba divers. Similar quadrats with floats attached are used for emergent or floating vegetation. In use, quadrats are placed at random in the area being sampled, and the plants within each are counted, or removed for further analysis.

A simple plant grab (Figure 5.7b) gives some idea of the species in a water body in a rather haphazard way, but its use is to be avoided if possible. Direct observation in shallow water and by scuba divers in deeper water is much more reliable, even qualitatively. Underwater photography is useful for recording macrophyte communities quantitatively with little disturbance, and the same spot may be photographed at intervals during the growing season—this is particularly easy with emergent and floating species (Figure 5.8).

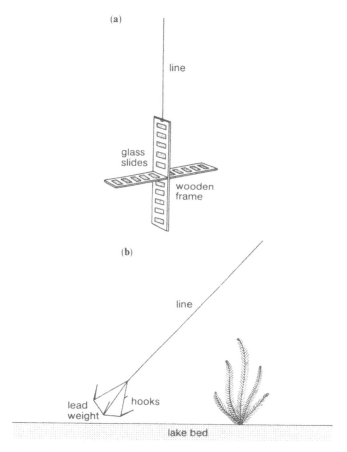

Figure 5.7 (a) Tray of slides for attached algae, and (b) plant grab.

Small macrophytes forming a sward can be measured quantitatively using the zoobenthos grabs and corers described below. In shallow water a broad corer on a pole is useful, being able to cut through the plant roots by pressure and rotation from above. In addition, where clear plastic core tubes are used, the sample is immediately visible on removal from the water, and the form and position of the plants seen *in situ*. In deeper water where the substrate is soft, a weighted Ekman grab samples some species satisfactorily.

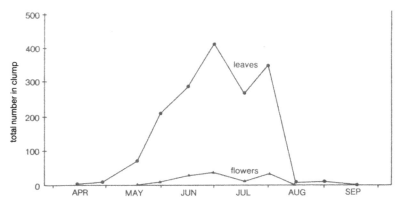

Figure 5.8 Direct counts of leaves and flowers within a single clump of water lily (*Nuphar*).

5.3.5 *Plankton*

It is often possible to use identical methods of sampling for both plant and animal plankton. If the population is homogeneous, only a few samples may be needed, but if the population is variable then many samples or mobile (towed) samples are necessary. Simple qualitative samples can be taken with a plankton net (Figure 5.9a) towed through the water, either vertically or horizontally. This net should have a fine mesh (70 meshes/cm) for phytoplankton, but a coarse mesh (20 meshes/cm) for large zoo-plankton (e.g. adult crustaceans).

Some of the best quantitative methods for collecting plankton involve apparatus already mentioned for water sampling (e.g. Friedinger, Ruttner, and van Dorn samplers). A simple stoppered bottle is useful to depths of about 15 m; all such bottles give known volumes of water from known depths and are widely used in plankton work. Where small quantities of phytoplankton from a standard column of water are required, a weighted tube may be used. It too will not function in water which is too deep. A pump may be used in conjunction with such a tube to collect samples from particular depths; these pumps are good for sampling phytoplankton, but less so for zooplankton, which may swim away from the mouth of the tube.

In many situations the numbers of certain plankton species are low, and standard water samplers do not collect a large enough volume. Several samplers have been evolved for such situations, most of them based on a

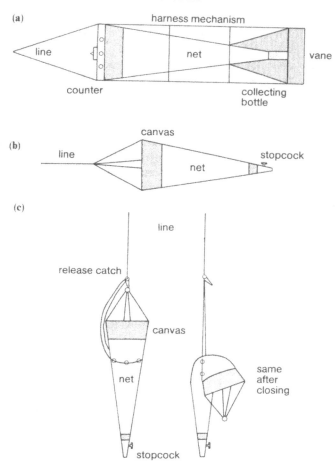

Figure 5.9 Plankton sampling: (a) Modern net with current meter and vane, (b) net, and (c) closing net.

net. At its simplest, this takes the form of a closing net which is hauled (usually vertically) at a constant rate through a known distance (Figure 5.9b). A triggering mechanism closes the mouth of the net at any point so that it can be removed from the water and emptied (Figure 5.9c). It is uncertain how much water such nets actually filter, and the more sophisticated samplers developed by Clarke and Bumpus (1950) and Slack (1955) measure the volume of water filtered by means of a meter.

The sophisticated horizontal sampler designed by Hardy (1959) and widely used for marine plankton can be used in fresh water. This sampler gives a quantitative record of zooplankton at a standard depth and, since the sample is collected on muslin and rolled continuously into a bath of preservative, the horizontal variation along the line sampled can be determined. A variety of other plankton samplers has been developed for specialised situations; these are not dealt with here.

Most plankton samplers have been used in lakes, but can be used also in running waters. Only in the lower reaches of rivers are true plankton important in running water; in smaller rivers and streams, however, bottom invertebrates occur in open water for considerable periods as 'stream drift'. This phenomenon has been studied by several workers (e.g. Muller, 1954), who have employed what are essentially large plankton nets held against the current; these nets usually have some device at the mouth to measure the amount of water filtered.

5.3.6 Benthic invertebrates

Because of the variation in bottom materials found in fresh waters, a range of equipment has been developed to sample these substrates and their invertebrates. Samplers used in standing waters are useful for the same substrates in running water and vice versa, except for stream samplers dependent on current to wash organisms into them. Soft homogeneous substrates are much easier to sample quantitatively than hard irregular ones.

For the qualitative sampling of bottom invertebrates many methods have been used—these have mainly involved open dredges in deep water, and dredges and hand nets in shallow water. In recent years, the quasi-quantitative technique of using a hand net in a standard way for a standard period of time has become very popular. The method has many advantages, especially where comparable collections are required from similar substrates which are difficult to sample by conventional quantitative 'area' methods.

Several grabs have been evolved for the quantitative sampling of finer sediments. These function by enclosing a standard area (usually about $200-300\,cm^2$) within the sampler walls. The jaws (normally released by a triggering mechanism operated from the surface) close underneath the sediment and the grab is raised to the surface. For mud, the Ekman grab—on either a pole or a rope—is one of the most popular instruments, but it is less useful in more solid substrates such as sand and fine gravels. Here the Petersen and van Veen grabs are commonly used.

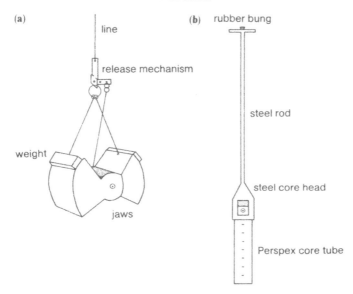

Figure 5.10 Bottom sampling, soft sediments: (a) Petersen grab and (b) Maitland corer.

Several different coring devices sample finer sediment, and these have advantages over grabs. The transparency of the tubes used means that the sample can be seen and its depth and intactness assessed. Most corers consist of a cylinder of some kind forced into the substrate by the weight of the sampler (Figure 5.10a) or by pressure on a pole attachment (Figure 5.10b). Some corers enclose the sample completely by means of flaps (e.g. Jenkin corer), before it is lifted from the bottom; others rely simply on sufficient adhesion between the sample and the wall of the corer to prevent sediment falling through the open bottom. Efficient corers are probably among the most accurate and reliable of bottom-sampling instruments.

Coarse gravels and stones cannot be sampled adequately by either grabs or corers. In running waters use is made of the current in the Surber sampler (Figure 5.11a)—a fixed net placed on the stream bed facing upstream. A standard area of substrate in front of the net is delineated by a quadrat and the material within this is lifted directly into the net or cleaned in front of it so that the current washes loose materials into the net. The shovel sampler, essentially a net with a shovel-like leading edge, functions by being pushed upstream through the substrate at a constant depth for a standard distance. By this means a known area of substrate

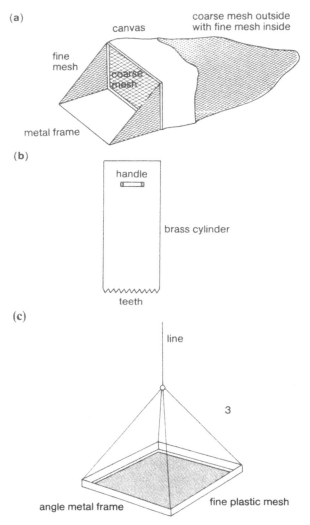

(a) canvas
coarse mesh outside
with fine mesh inside
fine mesh
coarse mesh
metal frame

(b) handle
brass cylinder
teeth

(c) line
3
angle metal frame
fine plastic mesh

Figure 5.11 Bottom sampling, hard sediments: (a) Surber sampler, (b) Wilding tube and (c) Moon tray.

is collected into the net. The shovel sampler can be used in standing water but is more efficient in running water where the current washes anything disturbed by the sampler into the net. A more useful sampler for stony lake shores is that of Wilding (Welch, 1948). Here, a large open cylinder with a toothed edge (Figure 5.11b) is forced into the substrate to a standard depth. All the coarse materials within the cylinder are then removed by hand, a bottom is fitted into the cylinder and the rest of the sample is lifted from the lake and poured into a container. A somewhat similar type of sampler was developed for stony running waters by Neill (Macan, 1958a), but here, after coarse materials have been removed, use is made of the current by opening a door on the upstream side of the cylinder and allowing the current to wash the animals in the cylinder into a net fixed to the downstream side. The Neill sampler is sometimes erroneously called the Aston sampler.

Invertebrates living among plant growths must be sampled by techniques related to the plant concerned and the substrate in which it is growing. Short plants in fine sediments can be sampled accurately by grabs or corers, but taller plants are more difficult to assess quantitatively. In many cases it is simplest to collect known volumes or weights of plants and find the numbers of animals in these. The relationship between plants and area can be estimated separately by one of the methods outlined above for macrophytes. Where animals are not washed off during the process, it is possible to cut out entire clumps of macrophytes into a net; this is most efficiently carried out in running waters, where the current washes loose materials into the net.

Benthic invertebrates have also been sampled using substitute substrates, which are placed in the study area and left for animals to colonise. Gauze trays were used by Moon (1935), filled with natural substrate, and then left *in situ* for standard periods of time before being lifted and examined (Figure 5.11c). Tiles, bricks or pieces of concrete of known area have been used by other workers. As with algae, the main problem is the artificiality of the situation presented to the animals, and the method, though useful, must be used with caution.

In many freshwater communities the dominant species are insects which emerge from water to air at well-defined periods of the year. Such species can be sampled accurately by emergence traps, which trap either the pupae under water as they rise to the surface or the adults at the surface as they emerge (Figure 5.12). Two types of sampler have been developed for standing waters: the funnel type, used mainly under water for pupae and adults, and the cage type floating on the surface for adults. In some

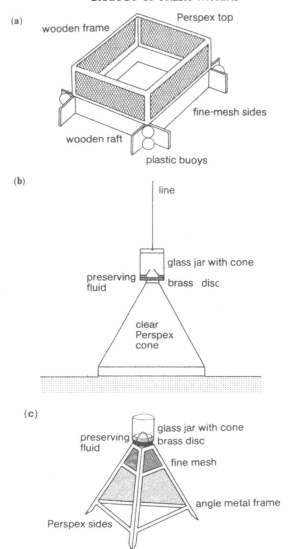

(a)

wooden frame   Perspex top

fine-mesh sides

wooden raft

plastic buoys

(b)

line

glass jar with cone

preserving
fluid   brass disc

clear
Perspex
cone

(c)

glass jar with cone
preserving   brass disc
fluid
fine mesh

angle metal frame

Perspex sides

Figure 5.12   Insect emergence traps: (a) box, (b) cone, and (c) stream.

slow-flowing rivers both these types can be successful, but where the current is fast a special design of trap must be used; usually this is based on the funnel type and fixed to the bottom.

5.3.7 *Fish*

Though fishing has been carried out by humans for many centuries, and several old methods are still in use, many of the most efficient methods of capture are based on relatively recent developments; notable among such are those based on electronics. Regardless of the basic method used to trap the fish, they are usually removed from the water by some form of net. A large number of methods is available for various species and situations; only the most useful general types are described here.

Nets are widely used to capture fish in fresh waters—especially in standing waters; the major kinds in use are seine, trawl and gill nets. The first two of these depend on trapping the fish by the movement of the net, while the last relies on the movement of the fish themselves into the net.

Seine nets (Figure 5.13) function by paying out a wall of netting from a boat so as to surround a certain area of water, enclosing the fish therein; the method is a popular and successful one. In shore seining, the net is set in a semi-circle from a small boat and hauled to the shore. Shore seine nets may or may not have a bag in the centre to help to retain fish. Most open-water seines do have such a bag, and the net is set from a boat in a full circle before hauling; two boats are often necessary. A rather specialised type of seine is that known as a purse seine or ring net; this is set, like the open-water seine net, in a full circle. It is deeper than other seines and along the bottom, running through large rings, is the purse line; as the net is hauled in, this line is tightened so as to close the bottom of the net.

Trawl nets are open bags of netting hauled through the water at constant speed by a powered boat. The main features of such nets are the shape

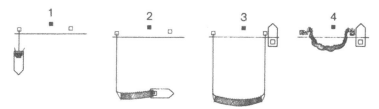

Figure 5.13 Operating a seine net from the shore.

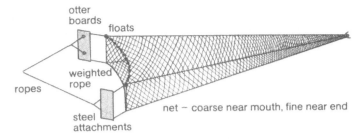

Figure 5.14   An otter trawl in action.

and length of the bag, and the method used to hold the mouth open. Two main types of trawl are in common use in fresh waters: beam trawls, where one or more rigid poles keep the mouth of the net open, and otter trawls (Figure 5.14), where planing surfaces known as otter boards keep the net open at either side, while floats and a weighted line keep it open in the middle. The bags of such nets must be sufficiently long to ensure that fish, having been overtaken by the mouth and moved back into the net, find it difficult to make their way out again.

Gill nets are sheets of netting hung in the water by an appropriate arrangement of floats, leaded lines and anchors (Figure 5.15). Since such nets depend on fish actually moving into them, they are made of fine

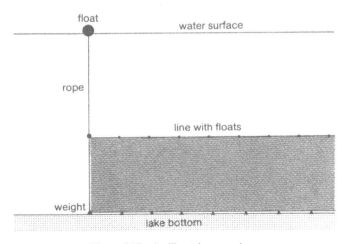

Figure 5.15   A gill net in operation.

materials coloured appropriate to the water in which they are being fished. Many of them are most successful in the dark or where visibility is poor. Gill nets depend on the fish swimming partly through an individual hole in the mesh and then becoming trapped—in many cases simply by the meshes slipping behind the gill covers so that the fish can move neither backwards nor forwards. In other cases the fish may continue to attempt to swim forwards so that its body becomes securely wedged in the meshes. The size of the meshes used in each net is critical in relation to the size of fish caught, and it is normal practice to use a range of mesh sizes in any nets fished for survey purposes.

Several other nets are used in specialised situations: large, coarse-meshed plankton nets (essentially trawls) are often used for small pelagic fish, while simple hand nets are very efficient in small waters, especially streams. Hand nets of various types are also widely used in other types of fishing as a method of retrieval of fish, e.g. in electric fishing or purse seining.

Trap nets can be extremely successful in certain situations. Many are relatively mobile and can be moved about easily; others are fixed more or less permanently to the bed of a lake or river. Most traps work on the general principle of fish moving, either by chance or because they are attracted in some way, through a cone or V-shaped entrance into a chamber from which they have difficulty in finding their way out. Many traps have two such chambers connected by a second funnel-shaped partition—this minimises the chances of fish ever finding their way out. Some traps have long sheets of netting which are used as lead nets to guide fish.

Almost all nets and traps depend on netting: formerly this was made of natural fibres, but these have been replaced for the most part by synthetic materials which are lighter, stronger and more durable. Many nets and traps are highly selective; this selectivity depends on the fishes' behaviour, the nature of the catching method, and the size of mesh used. Mesh size is often critical, and clearly must always be small enough to retain the smallest fish which it is hoped to catch. Its importance in relation to gill nets has already been mentioned.

Electricity is becoming more and more useful in connection with the capture of fish. Simple electric fishing gear, which is portable, produces a current which is passed into the water between two submerged electrodes. The electric field thus produced stimulates most fish within its area. With currents of the appropriate strength, where the current is alternating, the fish are stunned within its field; where the current is direct, the fish move towards the anode and are then stunned. In both cases the stunned fish are lifted easily from the water by a hand net; usually they recover quickly and can be released after examination.

Electric fishing is important as a method of capturing fish in running and some standing waters, and it can be one of the least selective methods of capture. Selection of the right type of gear is all important, however, as are appropriate safety precautions, for electric currents used in association with water can, of course, be lethal to humans. Apparatus using alternating currents tends to be easily portable and is useful in clear water where there is little cover for fish. Direct-current equipment, on the other hand, though heavier, is useful in turbid water or where there is a significant amount of cover. If the current used is too powerful, it may kill fish, while if it is too weak it may stimulate but not stun them, thus frightening them away. Weak currents are useful in this respect in forming barriers to lead fish into a trap or fish ladder, and recently so-called electric seines have been developed on this principle.

Though not a method of capture, the use of echo sounders and other electronic devices has recently become important in freshwater fishery work. Many very sensitive echo sounders are now available and are useful for indicating the position of shoals or even individual fish; nets can subsequently be fished in appropriate places to catch these. Echo sounding apparatus is also useful to study fish behaviour. It can be used either from a moving boat to study the numbers and dispersion of fish in a lake at any one time or from a stationary boat, to study the movement and activity of fish below. One of the main problems in using echo sounding for work of this kind is that of identifying the species concerned; this can often be overcome, however, by netting, underwater photography and explosives.

A variety of other methods is available for the study or capture of fish, but most are useful only in specialised situations, and are not of general application. The study of fish behaviour in the field may be carried out by direct observation in shallow clear water, or by the use of sub-aqua techniques, photography and television. Lesser methods of capture include the use of spears, gaffing, angling, explosives and poisons. Some fish poisons such as Rotenone (a derivative of derris) are extremely effective in clearing fish from waters for total population assessments or fishery management purposes. Rotenone acts initially as an anaesthetic and, if fish affected by it are placed immediately into fresh water, they normally recover. Certain other substances, notably MS_{222}, are used regularly in fishery research to anaesthetise fish during investigations—for example, where tags are being fitted.

CHAPTER SIX

ADAPTATION TO ENVIRONMENT: STRATEGIES FOR SURVIVAL

Of the three major environments found in the world today, fresh water is intermediate in character between sea and land—the former being stable and relatively uniform, the latter spanning a diverse range of habitats, many of which have fluctuating seasonal conditions. Though similar to the sea in being aquatic, freshwater habitats demonstrate a wide range of character, comparable to that found on land. In terms of water chemistry, inland waters range from those with practically no salts to brine lakes, in temperature from permanently frozen lakes to hot springs, in stability from the profundal regions of large lakes (where conditions are constant) to temporary pools (which dry up each year), and so on. In all these habitats 'freshwater' organisms exist and thrive. The present chapter examines the conditions in major habitats and illustrates how organisms are adapted to them. The examples cited highlight the major adaptations and should not be considered a comprehensive coverage of the enormous variety of form and function found among freshwater organisms.

The origin and evolution of organisms in fresh waters relates closely to their adaptations to that environment. Few freshwater organisms originated in their present medium; most are secondarily adapted to it, having evolved originally in the sea or on land. Some groups are found only in the sea: Foraminifera, Echinodermata, Cephalopoda, Branchiopoda, Cirripedia, Cumacea, Stomatopoda, Tunicata, Cephalochordata, Chaetognatha and other smaller groups. The Phaeophyceae, Rhodophyceae, Radiolaria, Porifera, Coelenterata, Polychaeta, Nemertea, Elasmobranchii and some other groups are found mainly in the sea with only a few species elsewhere. In fresh waters, on the other hand, Pteridophyta, Bryophyta, Dicotyledones, Anostraca, Notostraca, Conchostraca, Syncarida and Amphibia are common, but virtually absent from the sea. Monocotyledones, Rotifera, Nematomorpha, Cladocera, Hydracarina and Insecta are abundant in fresh water but comparatively rare in the sea.

Many groups common in fresh water also have large numbers of terrestrial species, and most evidence suggests the aquatic forms are secondarily adapted to this medium. Among such groups are the majority of aquatic Monocotyledones, Dicotyledones, Pulmonata, Insecta, Arachnida, Reptilia, Aves and Mammalia. Many aquatic members of these groups are actually amphibious rather than fully aquatic, or alternatively are terrestrial at some stage in their life cycle. Several groups are exclusively terrestrial, e.g. Onychophora, Chilopoda and most Arachnida.

Freshwater habitats undoubtedly impose restraints on life there, and many organisms have, by convergent evolution, developed similar adaptations. In contrast, others have evolved quite different mechanisms to cope with the same problems. Adaptations are of many kinds; in the account below, three main categories are considered: morphological, physiological and behavioural. Though an adaptation involving one of these may be of paramount importance in maintaining an organism in a particular environment, usually there is some involvement of all three categories to fulfil the requirements of a given niche.

Thus, though a few organisms have a long evolutionary history in fresh water, others have secondarily adapted from land or sea. Invasion from the land seems straightforward, and many Dicotyledones, Aves and Mammalia are intermediate by being amphibious. Invasion from the sea has arisen in two distinct ways: (a) by direct migration via estuaries or brackish waters (e.g. the Baltic Sea) and (b) through isolation, mainly by ice action and movements of the earth's crust, and subsequent freshening of former seas (e.g. Lake Baikal). The adaptations involved in both cases are essentially similar.

6.1 Major adaptations

6.1.1 *Shape*

The shape of many aquatic plants and animals is related to the relatively high density of water compared to air and especially to the extent to which the various species have either to resist water currents or to move through the water. It is often possible to describe the nature of the habitat occupied by an animal simply from its general shape. There is less modification in sessile forms than in motile ones but even in the former, lotic forms are more streamlined than lentic ones.

Numerous experiments related both to biology and to human

navigation have shown that a fusiform shape offers least resistance to the aquatic medium and the ideal streamlined body is widest at about 36% from its pointed anterior end tapering to a point at the rear. This is the commonest general shape found among boats and in most of the best animal swimmers in the world (i.e. certain fish and aquatic mammals) and is also found among those invertebrates (e.g. some mayfly larvae) which are exposed to strong currents (e.g. *Baetis*) or can swim (e.g. *Leptophlebia*).

Motility is unimportant to higher plants but lotic species do have to resist strong currents and such species have a number of features in common. These include strong attachment to the substrate, very flexible stems (which will bend, thus offering little resistance to the current) and highly dissected, linear or streamlined leaves.

Many invertebrates and some fish, where mobility or efficient resistance to currents are not important, show relatively little streamlining and the body form may be modified in some other way, relevant to their life styles. Thus animals which live in burrows (e.g. worms and midge larvae) are usually thin, elongate and poor swimmers.

Animals which live on or under stones show various degrees of streamlining, depending again on their motility and/or the strength of the current, but very many of these exhibit a high degree of flattening. This has two important advantages. Firstly it allows them to press close to the substrate and so take advantage of the relatively still boundary layer there. Secondly, it allows them to move into crevices and other small spaces and so escape the current (as well as larger predators). Most flatworms and many mayflies come into this category and these and other streamlined flattened forms are among the commonest and most abundant of lotic animals.

Most of these plants and animals where resistance to the current is important have any projections (antennae, etc.) reduced to the minimum. In contrast, the small plants and animals of the phytoplankton, where buoyancy is all important in maintaining vertical position in the water column, often have various projections (e.g. long setae or antennae) to reduce their sinking rate.

6.1.2 *Osmoregulation*

A major problem encountered by organisms invading fresh waters from the sea is the difference in osmotic conditions between the two environments. In freshwater organisms, body fluids are hypertonic to the water (usually low in most salts) which continually passes into the body through permeable surfaces. The areas of permeable tissue are as reduced as possible, many surfaces being covered by impermeable chitin or cuticle. The most permeable areas are normally the respiratory tissues—gills and lungs. Marine animals die in fresh water because they are unable to control the rapid entry of fresh water which dilutes internal fluids causing swelling and death.

In most higher aquatic plants the developing leaves are submerged (Sculthorpe, 1967). Such leaves are sites of high metabolic activity, where sugars are synthesised and osmotically active solutes translocated from other parts of the plant; because of this they possess a high osmotic pressure. Most young shoots are protected from the inward passage of water by a mucilaginous sheath or by early cuticularisation; in adult plants most of the epidermis is cuticularised.

Most freshwater invertebrates have developed efficient urinary systems involving, for example, contractile vacuoles or nephridia—enabling them to produce quantities of dilute urine which can be excreted, thus removing excess water and maintaining internal stability. This urine, though normally dilute, does contain salts; these are either selectively reabsorbed or replaced by others absorbed from food. Vertebrates other than fish rarely have exposed permeable tissues. Freshwater fish, which do, have well-adapted kidneys and excrete dilute urine. Marine fish, in contrast, lose water to the sea (since their body fluids are hypertonic to it); they counteract this by drinking water and excreting concentrated urine. They can also excrete excess salts via specialised gill cells. Organisms which can thrive only in a narrow salinity range (high or low) are known as stenohaline; those tolerating widely fluctuating salinities (as found in estuaries) are euryhaline.

Closely related animals living in the sea and in fresh water show marked physiological differences. Freshwater forms use more oxygen than their marine counterparts, due to greater energy expenditure in maintaining constant internal osmotic pressure. Contractile vacuoles, the main method of excretion in freshwater protozoans, are unimportant in marine forms. In many freshwater arthropods, kidneys are better developed than in related marine species. Among fish encountering the greatest ranges of salinity during their life cycles, and coping with associated osmoregulatory problems, are diadromous species (Figure 6.1). The sea lamprey *Petromyzon marinus* and the Atlantic salmon *Salmo salar* are anadromous, starting life in fresh water, passing to the sea to mature, and then returning to fresh water to spawn.

The European eel *Anguilla anguilla*, on the other hand, is catadromous, spawning in the sea but moving to fresh water and growing there before

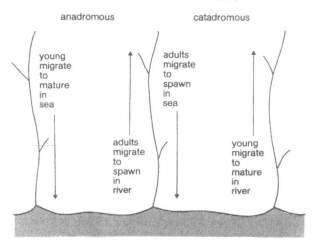

Figure 6.1   The life cycle of anadromous (e.g. salmon) and catadromous (e.g. eel) fish.

returning to salt water to spawn. Such fish are extremely adaptable in their ability to control osmosis, and have glomerular kidneys which can adjust urine volumes according to different salinities encountered. These adjustments are controlled by other physiological activities, particularly those of certain endocrine glands.

Physiological adaptations may combine with ecological ones to prevent genetic breakdown in some species (Lagler *et al.*, 1962). In parts of Europe the three-spined stickleback *Gasterosteus aculeatus* occurs as two populations, distinct physiologically and ecologically; one population lives permanently in the sea (migrating into estuaries to spawn), the other lives permanently in fresh water. The two races are distinguishable morphologically, though many intermediate types are found. Physiologically, however, the two races are separate, and no intermediates exist. The saline form has difficulty controlling blood chloride in fresh or low-salinity water, and may die there. In contrast, the freshwater form has the osmoregulatory capacity to maintain its blood chloride, even in pure water. The two races, therefore, rarely meet, and genetic mixing is prevented.

6.1.3   *Respiration*

Gas exchange is common to all freshwater organisms, but related adaptations vary considerably according to origin, size, activity and habitat in which an organism lives. Those originating in the sea, though encountering osmoregulatory problems, are well adapted for aquatic respiration in fresh water. Animals originating from land, on the other hand, though partly pre-adapted to osmoregulation, encounter gas exchange problems under water and have evolved various mechanisms to deal with this.

In unicellular organisms, gas exchange is effected by simple diffusion through the body wall. In higher plants the permeability of the thin cuticle and epidermis to dissolved gases aids gas exchange, while the extensive lacunae within the tissues, characteristic of aquatic species, facilitate transport of gases internally. Of invertebrates which have invaded fresh water from land, insects are the most successful. They have overcome the problem of gas exchange under water in two ways. Some are completely aquatic and permeable areas of cuticle serve for gas exchange. These species are widespread in fresh waters, but only move from one to another during the adult aerial phase. Other groups respire through a bubble of air attached to hair on the body surface or under the wing covers. This bubble is periodically renewed at the water surface, and such insects, though mobile and moving easily from one water to another, are restricted to shallow and still waters where depth and current do not interfere with movement to and from the surface (Figure 6.2).

A similar situation is found in pulmonate molluscs which respire through a lung developed from the mantle cavity. In some this lung is filled with water and they are independent of depth; in others it is air-filled and the animals can survive only in shallow water where they surface from time to time to breathe. Populations of one species (e.g. the wandering

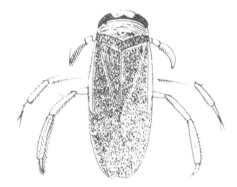

Figure 6.2   An air-breathing aquatic insect: *Corixa.*

snail *Lymnaea pereger*) are known to exist in the same water using both methods: one group in deeper water using aqueous lungs, the other in shallow water with aerial lungs.

Fish are the only vertebrates to have developed (in their filamentous gills) an extremely efficient means of aqueous respiration. Though most amphibians (especially the larval stages) can breathe under water, freshwater reptiles, birds and mammals must periodically rise to the surface to renew air, and are restricted in the freshwater habitats which they can utilise efficiently.

6.1.4 *Other physiological features*

Colour range among freshwater organisms varies from the complete transparency of plankton to the varied coloration of benthic invertebrates and most vertebrates. The colour of plants is dominated by their respiratory pigments. Colour in animals is mainly adaptive (Cott, 1940), and its functions are threefold: advertisement, camouflage and disguise. The brightest and greatest range of colours among aquatic animals are in various fish (especially coral reef species), where the colours are due to cells known as iridocytes and chromatophores. Many fish can change colour rapidly according to background or physiological state, or may do so gradually at different stages in their life history.

Several fish are unique in being able to generate electrical discharges; these currents are produced by special tissues—usually modified from muscle or gland. In *Gymnotus* the large electric eel of South American rivers, much of the posterior muscle is modified into an electric organ capable of generating a strong stunning current, while in *Malapterurus*, the electric catfish of Africa, a similar high voltage is produced by modified skin glands. The currents are used to stun prey so that they can be caught easily. Some fish emit continuous weak electric pulses which enable them (with electrical receptors) to detect objects in turbid water or during darkness.

6.1.5 *Buoyancy*

An important adaptation in aquatic organisms is the development of a hydrostatic system, aided by a particular shape, to slow down sinking. This means that such organisms waste little energy in maintaining themselves vertically. Most fish have a gas-filled swim bladder whose volume can be adjusted to any pressure so that the fish is neutrally buoyant.

Many algae and some invertebrates have gas bubbles to make them more buoyant. Higher aquatic plants have a reduced skeleton and are unable to support themselves in air. In water, however, they are supported by gas in lacunate tissues, which gives them such buoyancy (Sculthorpe, 1967) that stems, not having to support the weight of leaves, often attain great lengths or can branch freely as they grow.

Many methods of attachment to the substrate are found among sessile animals, mainly suckers or adhesive mechanisms. In motile forms, many invertebrates and most fish are efficient swimmers, characterised by a streamlined body, a propulsive device (often the tail or trunk region, sometimes the limbs) and stabilising fins or setae. The swimming limbs of arthropods have well-developed fringes of setae to aid propulsion and mobility in the water.

6.1.6 Symbiosis

In symbiosis, two or more different organisms live in close association with one another and are often highly adapted morphologically, physiologically or behaviourally to this existence. There are several types of symbiosis: parasitism, where one of a pair thrives at the expense of the other; commensalism, where one member benefits from the association but does not notably affect the other; and mutualism, where both partners benefit from the association. Symbiotic relationships are found in all environments and the few examples cited below are intended to demonstrate some of the inter-organic relationships and adaptations which occur in fresh water.

Though parasites may damage their hosts, they rarely kill them, since this is normally to their own disadvantage. Many aquatic ectoparasites have behavioural adaptations to find their hosts but most important modifications are morphological—usually elaborate attachment organs (hooks or suckers) and complex sucking mouthparts. In the fish louse *Argulus* the body is flattened so that, when attached to a fish, it offers little resistance during swimming. Attachment is by two suckers which adhere to the host's skin; feeding is by piercing mouthparts which penetrate the skin and suck out body fluids. These parasites can leave their hosts for considerable periods, swimming from one host to another or finding a substrate on which to lay eggs. In endoparasites, on the other hand (e.g. tapeworms), the animals are completely adapted to life within the host; there is little difference in form between the cestodes found in aquatic and terrestrial hosts—all have well-developed hooks or suckers for attachment and are physiologically adapted for respiration and feeding within the host's body.

Many algae are highly adapted to a symbiotic existence with other organisms including bacteria, fungi, spermatophytes and vertebrates. Several examples have been discussed above (e.g. algae and fungi combining to form lichens; *Anabaena* with *Azolla*). Tiffany (1951) points out the close relationship between two algae, the combined form described as *Glaucocystis*. In fact, its chloroplasts are blue–green algae (now assigned to the Chlorococcales), living inside a colourless green alga belonging to the Chlorophyceae. Similar close relationships may be found between certain Myxophyceae and nitrogen-fixing bacteria such as *Azotobacter*.

Mutualism occurs in the partnership between the bitterling *Rhodeus* and the mussel *Anodonta* (Figure 6.3)—both adapted morphologically and behaviourally for this relationship. The bitterling is a small cyprinid fish unusual only in that during breeding the female develops a long ovipositor. When spawning, the male and female select a suitable mussel and swim over it, the female passing eggs into the mantle cavity by means of its ovipositor, the male's sperm being carried in by the inhalent current. The few large eggs are protected by the mussel until after hatching, when the fry leave via the exhalent siphon. During reproduction of the mussel, its

Figure 6.3 Bitterling (*Rhodeus*) spawning in a mussel (*Anodonta*).

eggs remain inside the mantle cavity until they hatch into larvae known as glochidia. When bitterling (or other fish) pass the mussel these glochidia are ejected by contraction of the shell valves, and the larvae attach to fish for a short period, after which they drop off and develop into normal bivalves. The main function of this stage is one of dispersal.

6.1.7 *Behaviour*

Behaviour is broadly defined as action which alters the relationship between an organism and its environment (Kimball, 1965). The behavioural responses of freshwater organisms are adaptations to their life there, but closely related forms may demonstrate widely different behaviour for survival in various environments. Behaviour in plants is simple and innate. In higher plants it is restricted to growth movements dictated by light and temperature. In motile algae (e.g. *Chlamydomonas*) directional movement is possible—usually towards or away from light, so that they find themselves in that intensity best suited to their metabolism.

Invertebrates and vertebrates have developed more complex behaviour patterns, many of them innate, but in higher forms (e.g. fish and mammals) learning processes are involved. The importance of adapted behaviour patterns to the life of these animals in varied and complex environments is a subject too vast to be dealt with here other than by a few examples.

The behaviour of many animals, especially during breeding, is often relevant to their further distribution and success in a particular water. Macan (1963) demonstrated this with two similar damselfly larvae in a mountain tarn. Both species are found in shallow water there, except in thick *Carex* beds where only one (*Pyrrhosoma*) is numerous. In contrast, only the other (*Enallagma*) occurs in *Myriophylpum* beds in the middle of the tarn. The difference is due to the egg-laying behaviour of the females. *Enallagma* adults avoid thick emergent vegetation, but select individual plants (e.g. *Myriophyllum*) near the surface, climb down these and lay their eggs. *Pyrrhosoma* females, in contrast, alight on the floating and emergent leaves of plants (e.g. *Carex*), and lay their eggs into stems just below the surface. There is little movement of the larvae afterwards outside the area in which they have hatched.

The complex behaviour patterns of some fish at spawning are well known, and the individual responses of a few species, e.g. the three-spined stickleback *Gasterosteus aculeatus*, have been worked out in detail. In addition to such behavioural responses, many species are adapted towards

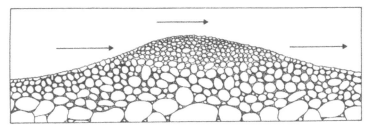

Figure 6.4 A characteristic salmonid spawning bed in a stream (after Stuart, 1953). Only at the downstream end of a pool is there sufficient current through the gravel to allow buried eggs to respire.

selecting a suitable spawning site before egg-laying commences. Though it was known for some time that female trout *Salmo trutta* dig out and spawn in depressions in gravels, it was some time before Stuart (1953) demonstrated that the gravels chosen do not occur in all parts of the stream but only in areas where the contour of the bed creates a flow of water through the gravel, allowing aeration of eggs after they have been laid and covered over by gravel (Figure 6.4).

6.1.8 *Dispersal*

An important aspect of the ecology of some freshwater organisms is their power of dispersal, and many species are adapted morphologically and behaviourally for this purpose; the distribution of the larvae of *Anodonta* has already been discussed. Lower organisms (bacteria, algae and protozoans) are often capable of encystment or can produce spores (e.g. fungi and mosses) which are dispersed by the wind and other agents. Most insects have high powers of dispersal during the adult aerial phase, but other groups (e.g. coelenterates, annelids, some crustaceans and many fish) are poorly adapted and restricted in their distribution.

Several aquatic macrophytes are capable of rapid growth from very small fragments which can be dispersed variously by floods, wildfowl or humans. Casual transport is probably significant in short-distance dispersal of plants (Sculthorpe, 1967) and animals (Hunter *et al.*, 1963). A dramatic example of the rapidity and efficiency of dispersal in a macrophyte is the spread of the Canadian pondweed *Elodea canadensis* (Figure 6.5) in Great Britain. Probably introduced into Europe via Ireland about 1836, it was first recorded in Great Britain in a pond at Duns in Scotland in 1842. By 1847, it had been recorded at several places in

Figure 6.5   An invasive alien to Europe: Canadian pond weed (*Elodea canadensis*).

England. Within a few years *Elodea canadensis* had invaded waters in most parts of the country (often via canal systems), blocking sluices and disrupting waterways. It also spread widely through the rest of Europe and more recently has become common in Australasia. Similar examples of dispersal are found in animals, e.g. *Potamopyrgus jenkinsi* (Figure 6.6), a small operculate snail, was first found in the British Isles in the Thames estuary in 1883 but has since been recorded in most parts of the country.

Figure 6.6   A recent invader of fresh waters in Britain: *Potamopyrgus jenkinsi.*

In some marine animals the eggs and larvae are planktonic, primarily for dispersal. During the evolution of freshwater forms from marine ones, however, such pelagic stages have been disadvantageous, because of the danger (especially in rivers) of being carried to the sea. For this reason, only a few freshwater organisms have pelagic larvae. Nevertheless many invertebrates have problems of dispersal within a water during the early life stages; in insects especially, the eggs are often laid at or near the water's edge, while the larvae live in profundal muds at considerable distances from the shore. The early larvae of many of these animals are positively phototactic and semi-planktonic; by this means they are able to drift in the water for some time until appropriate currents carry them to a suitable substrate on which they can settle.

6.2 Adaptation to specific habitats

6.2.1 *Running waters*

The communities of running waters are characteristic of this habitat and include forms restricted to and highly adapted for life there. The major factor affecting both the habitat and the ability of the organisms to live in running water systems is the current. A river bed is typically rocky or stony, and most organisms are adapted to the current either by withstanding its force (by some morphological adaptation) or by sheltering from it (usually by a behavioural response). In animals, the problem of feeding in strong currents has also produced many adaptations of form or behaviour.

Many algae are adapted to life in swift currents, exhibiting similar features to animals there, i.e. streamlining and attachment organs. Attachment in stream algae is by holdfast cells and with ensheathing peptic compounds. Two growth forms in streams have been distinguished: those attached only at the base, and forming long streamers in the current (e.g. *Cladophora*), and those which encrust and are completely adhesive (e.g. *Lithoderma*). The stems of higher aquatic plants are distinguished from those of terrestrial ones by a reduction of the skeletal lignified xylem tissue and the merging of vascular strands into a central cortex. Schenck (1886) suggests that the latter modification is adaptive; since aquatic plants (especially those in running water) have to endure pulling strains (not bending ones as in terrestrial forms), these are best withstood by having some form of rope-like core.

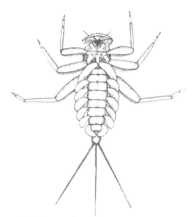

Figure 6.7　A ventral view of the larva of a stream mayfly (*Rhithrogena*), showing sucker-like gill formation.

Mayfly (Ephemeroptera) larvae are prominent components of stream faunas; these larvae are often streamlined in form and flattened to withstand the current. In some (e.g. *Rhithrogena*) the abdominal gills are developed in a fan on either side to form a sucker enabling the animal to cling to smooth surfaces (Figure 6.7).

Larvae and pupae of the blackfly *Simulium* are confined to running waters and are characteristic members of the lotic community. This is one of the few invertebrates which can tolerate exposure to torrential conditions on smooth surfaces, and the larvae are highly adapted for this mode of life. The adult females, which suck the blood of various birds and mammals, lay their eggs at the edge of running waters; after hatching, the larvae gradually distribute over the stream bed, attaching to stones and weed where the current is strong. The larvae are tubular and slightly swollen at either end, especially the posterior where radial rows of fine hooks form an organ of attachment (Figure 6.8). Anteriorly a modified proleg on the thorax ends in fine hooks as an additional means of attachment—especially important when the larvae move. The larvae also spin a fine thread to attach themselves to the substrate; in this way, even if displaced they can move back to an original position via this thread. When feeding, larvae stretch out into the current, holding on by the posterior sucker only. The mouthparts include a pair of comb-like sieves which filter material from the water and pass it periodically to the mouth. On pupation, the larvae move to a more sheltered area and spin a

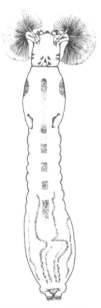

Figure 6.8 A larva of the blackfly (*Simulium*), a filter-feeder in streams.

slipper-shaped cocoon, the opening of which is positioned away from the current so that when the adult fly emerges it is carried to the water surface by the upward current.

Larvae of the caddisfly *Hydropsyche* are also found only in running waters where they live under stones or in weed beds in fast-flowing water. Eggs laid under water by the aerial adults hatch into larvae which find suitable positions in the stream bed. Here they construct loose silken shelters and nets (Figure 6.9) which have wide openings into the current, so that particulate organic material is filtered out by their fine meshes. The back of the net leads into the chamber containing the animal which periodically cleans food off the net and eats it.

6.2.2 *Profundal regions*

The profundal region, though often extremely stable, can present extreme conditions for organisms occurring there. The main features of the profundal habitat are the absence of light, the periodic depletion of oxygen (in eutrophic lakes) and the soft nature of the sediments, which

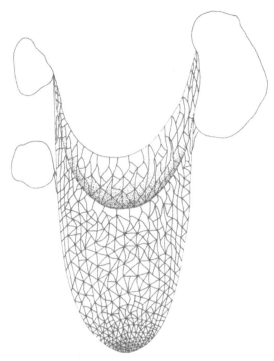

Figure 6.9   Silken net of the caddis larva *Hydropsyche*.

offer little in the way of cover or protection. Burrowing invertebrates are characteristic inhabitants of such areas, feeding on suspended or sedimented materials, and capable of tolerating low oxygen levels or even completely anaerobic conditions.

The immature stages of the midge *Chironomus* are characteristic inhabitants of profundal lake muds, where, during stratification, dense populations can tolerate anaerobic conditions. The adult midges lay eggs in jelly masses on the water surface, and when semi-planktonic larvae hatching from these drift over soft muds they burrow and build U-shaped tubes (Figure 6.10). The tube walls are held together by fine silk, while inside the tube each larva spins a fine silken net. By undulating rhythmically the larva draws water inside the tube, carrying with it oxygen and particulate food. This is filtered by the net, which is eaten periodically and then re-spun. Faecal pellets and other waste materials pass from the tube with the outgoing current. During low oxygen levels

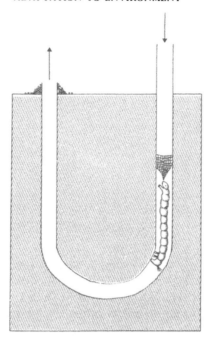

Figure 6.10 Life style of the larva of the midge *Chironomus* at the bottom of a lake.

haemoglobin in the blood of the larvae makes activity possible, but with completely anaerobic conditions activity ceases and the larvae become quiescent until oxygen is available again. Pupation is within the tube, but after a few days the pupa wriggles out and swims to the water surface (often aided by the buoyancy of gas which develops between the pupal and adult skins) where the adult emerges.

Other characteristic inhabitants of profundal muds (and other microhabitats where sediment is soft and oxygen concentrations low) are tubificid oligochaete worms. *Tubifex* itself is typical of the family and often abundant in profundal areas. In the soft sediment there, these worms build fine tubes which (like those of *Chironomus*) project from the surface (Figure 6.11). The posterior end of the animal projects from the end of the tube and is waved rhythmically to create local currents. Most Tubificidae are reddish due to a respiratory pigment in the blood. Food is ingested below the mud suface and consists of sedimented material, especially phytoplankton and dead invertebrates. Faeces pass out at the

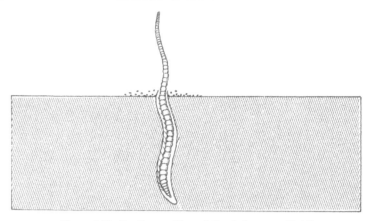

Figure 6.11 Feeding position of the sludgeworm *Tubifex*.

posterior tip and, due to the rhythmic waving process, are deposited at some distance from the tube opening. Any disturbance in the vicinity of the tube causes the exposed portion to be pulled into the safety of the tube. Even if the posterior end is bitten off by a fish, it is capable of regeneration.

Some organisms thrive only in situations devoid of oxygen; among these forms are the methane bacteria (e.g. *Methanocarcina*). This bacterium flourishes in the deep sediments of lake muds where there is no oxygen, and even hydrogen sulphide is absent. Under such conditions plant cellulose (previously partly broken down by other bacteria) is further broken down with the release of methane gas. This gas characterises anaerobic conditions and is released from the profundal muds of some lakes during thermal stratification or under complete ice cover.

6.2.3   *Plankton*

Freshwater plankton is a distinct, and relatively self-contained, community whose members exhibit common adaptations. Most planktonic species are small—the dominant groups being bacteria, algae, protozoans, rotifers and crustaceans. Small size (with a high surface to volume ratio) is related to a necessary adaptational feature of all plankton—the ability to prevent sinking. In only a few plankton species is the density the same as or less than that of water; in most cases density is greater, and slow sinking is inevitable. Positive buoyancy is achieved by gas bubbles or oil globules, while the slowing of sinking is achieved by adaptation of body form.

Planktonic organisms with skeletons (e.g. diatoms and rotifers) have lighter structures than related non-planktonic forms. Many organisms have gelatinous sheaths, or contain large volumes of water within them to reduce total density. Most animals are able to maintain their vertical position in the water by constant motion. Non-motile members of the plankton whose density is greater than water rely almost entirely on turbulent currents to keep themselves suspended there.

The most striking modifications among passive phytoplankton to aid their existence in the open-water community are flotation devices. Most planktonic algae have a large surface area relative to volume, so that increased friction reduces the rate of sinking. Many phytoplankton are therefore elongate. *Asterionella* is not only elongate but has become colonial, so that the long cells form a star shape which has a parachute-like effect to slow sinking (Figure 6.12). Another means of aiding flotation is the inclusion of vacuoles within the cell. *Anabaena*, like many other blue–green algae, has inclusions in the plasma which lower density so much that during calm weather individuals float to the surface and form a 'bloom'.

The transparent larvae and pupae of the phantom midge *Chaoborus* are confined to standing waters and are a characteristic component of many plankton communities. They are highly adapted for this mode of life. Spiral egg masses laid at the water surface by aerial adults hatch into small larvae which disperse throughout the water body, carrying out vertical circadian migrations, but sometimes remaining buried in the mud for part of each day. The larvae are carnivorous with prehensile antennae which seize other small planktonic organisms. The elongate cylindrical

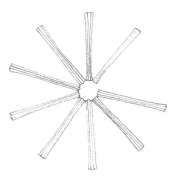

Figure 6.12   *Asterionella*, a characteristic member of the phytoplankton of many temperate lakes.

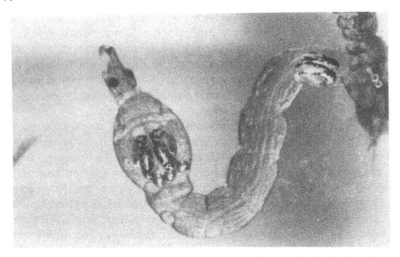

Figure 6.13   The planktonic larva of the phantom midge (*Chaoborus*), showing the pigmented hydrostatic organs used to maintain buoyancy (Photo: D.B.C. Scott.)

body, like that of other zooplankton, is almost completely transparent, apart from two pairs of darkened hydrostatic organs (Figure 6.13) which enable the animal to remain motionless in the water awaiting its prey. The last abdominal segment carries a comb of stiff setae which acts as an organ of propulsion (Figure 6.14). The transparent pupa has a large pair of anterior breathing organs, and posteriorly a pair of flexible paddles which propel it towards the surface where emergence takes place.

Another animal adapted to a planktonic existence is *Leptodora*, a predacious crustacean found in the open water of many lakes. Apart from its eyes, and material in the gut, it is completely transparent and very difficult to see. The carapace (prominent in most cladocerans) is absent and the abdomen elongate. The first antennae are large and used for swimming. The first pair of legs are modified to form clasping limbs to seize prey and hold it to the mouth, while the remaining legs are small (Figure 6.15). These animals feed by swimming through the water, periodically seizing and devouring small zooplankton (mainly other crustaceans and rotifers) which come within reach.

Figure 6.14   Transparent planktonic larva of the phantom midge *Chaoborus*.

Figure 6.15 One of the larger predatory zooplankton *Leptodora*.

6.2.4 *Neuston*

Unimportant in marine environments, the water surface of fresh waters are still or slow-flowing and has a characteristic community. The organisms concerned are highly adapted to this specialised aquatic niche at the air/water interface. The epineuston live above the water surface, the hyponeuston just below it. Some organisms are at times part of the neuston but can also descend to the depths below or fly into the air above.

The genus *Azolla* (Pteridophyta) is free-floating and occurs in abundance in small standing waters in many parts of the world—often covering the surface and dominating the habitat. The form of the plant exhibits extreme reduction. The stem branches pinnately and carries two alternating series of minute leaves, each divided into two parts (Figure 6.16). The submerged portion is colourless and absorbs water and nutrients; the roots are poorly developed as small hair-like structures. The aerial portion of the leaf

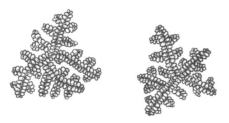

Figure 6.16 Two plants of the floating fairy moss *Azolla*.

contains the photosynthetic tissue, but also a cavity in which the blue–green alga *Anabaena azollae*, a symbiotic species, fixes atmospheric nitrogen. Buoyancy is achieved by air trapped between leaves and stem, and small epidermal hairs on the dorsal leaf surfaces which trap bubbles of air, keeping the plant dry and preventing immersion during rainfall.

The bladderwort, *Utricularia*, is a rootless, floating, submerged plant, of interest not only for its hyponeustic habit but also because it is carnivorous. The plant consists of an elongate branching axis with numerous finely dissected leaves; these bear bladder-like traps which are small bulbous structures with an opening which can be closed by a flap valve. Numerous hairs around this opening form a funnel leading into it. The outer surface of the flap bears stiff bristles acting as triggers for the closing mechanism, and also stalked glands which secrete sugar to attract prey. Glands inside the wall of the bladder extract water, thus reducing pressure; small aquatic organisms (e.g. protozoans and rotifers) touching the trigger hairs cause the flap to open, and water and prey are sucked inside (Figure 6.17). The prey dies and breaks down, releasing soluble products then absorbed by the plant. The mechanism is an adaptation to life in habitats deficient in nitrogen; *Utricularia* is exceptional among free-floating plants in being able to colonise fen and bog waters with very low pH and salt concentrations (Sculthorpe, 1967).

Active members of the neuston include the water skater *Gerris*, which is one of the main predators found among this community. Both nymphs and adults are capable of moving rapidly over the water surface by a series of quick strides. Sensory organs are well developed, and there is a pair of large eyes and long antennae; as in many Hemiptera, the mandibles form an elongate piercing proboscis. The three pairs of legs are also elongate, the middle pair being held well out at the sides; these are the

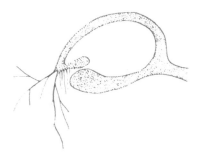

Figure 6.17   Section of the feeding mechanism of the bladderwort *Utricularia*.

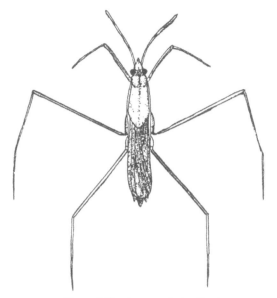

Figure 6.18   A water skater *Gerris.*

means by which *Gerris* is able to make long leaping movements
(Figure 6.18). The body is clothed in fine velvety hairs at right angles to
the integument; these are water-resistant, and the animal is able to stand
on the surface film by means of hair pads on the widely spread feet. Even
when accidentally submerged by a raindrop the animal is never wetted.
Gerris is found on many quiet lakes, ponds and slow-flowing rivers and
is extremely sensitive to surface vibrations. When a terrestrial insect falls
on to the surface and is trapped there, it is quickly seized and its body
fluids sucked by *Gerris.*

Adult whirligig beetles *Gyrinus* are highly adapted for a dual life at the
water surface and below. The eggs are laid on submerged vegetation and
hatch into elongate larvae with long filamentous gills; this stage is
completely aquatic and part of the weed benthos. After pupation, the
adults emerge and spend much of their time gliding swiftly on the surface
film feeding on terrestrial invertebrates trapped there. When alarmed or
during poor weather they submerge, holding on to weeds or stones by
their hooked forelegs. Adult beetles are black, shiny and oval; the middle
and hind legs are modified for swimming, being short but flattened and
paddle-like with a well-developed fringe of hairs. The compound eyes are

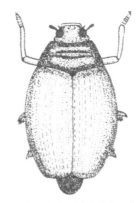

Figure 6.19   An adult whirligig beetle *Gyrinus*.

divided by the antennae (which are short and streamlined) so there are
actually two pairs—the upper pair for aerial vision, the lower pair for
under water (Figure 6.19). When diving, like many other beetles, they
carry air under the elytra for respiration.

6.2.5 *Subterranean aquatic systems*

Much has been learned in recent years about the ecology of underground
waters, which are much more common on a world-wide basis than had
been realised previously. Within the spectrum of freshwater ecosystems
they have several unique features which make them of particular interest
to ecologists. These include a total absence of green plants, and in the
animals, many specific adaptations including a loss of periodicity in
reproduction, an absence of pigments, virtual blindness and the
development of various tactile organs.

The surface of water held underground in permeable rock strata is
called the water table. Water below this is called ground water, water
above this (usually draining through soil and rocks) is termed vadose
water. Areas of water held below ground under pressure are known as
artesian basins. These are important sources of human drinking water in
some countries.

Regardless of their geographic position, subterranean systems, which
include both rivers and lakes, have much more in common with each
other than comparable waters above ground. Typical environmental
conditions include a total absence of light, constant (usually rather low)

temperatures, an absence of primary producers and a consequent shortage of organic food for secondary producers. The spatial extent of these systems can be quite variable from isolated small spring-fed waters to extensive interconnected underground rivers and lakes.

There are two main categories of underground system: conventional standing and running waters of various sizes, and interstitial waters within gravels which, though they can be very extensive, are very limiting in a spatial context. Thus the animals within interstitial systems are all very small to enable them to move among the water-filled crevices in the gravels. Interstitial communities are much commoner than is realised and frequently interconnect (and share some species) with the gravel beds of conventional lakes and rivers.

The absence of green plants underground is only partly made up by the presence of colonies of bacteria and fungi which serve as a source of energy to secondary producers. The major source of food in most underground systems is allochthonous material (e.g. leaf particles) from above ground which is washed in by rain. The secondary invertebrate producers which feed on these materials are themselves eaten by a variety of carnivorous forms.

The small animals which commonly make up the communities of interstitial systems are mainly nematodes, tardigrades, oligochaetes, hydracarines and an extremely interesting variety of crustaceans (often endemic species) which include syncarids, copepods, isopods and amphipods. In the larger underground waters many more crustaceans appear including some prawns and crabs, as well as a variety of higher predators such as blind fish and amphibians.

6.2.6 *Temporary aquatic systems*

Special adaptations occur in many aquatic organisms towards a temporary drying-out of their environment or towards a circadian or seasonal change of habitat from water to air or vice versa. Many aquatic communities are dominated by insects, the adult stages of which may be independent of water; the adults of many species are merely a short-lived reproductive and dispersal phase which is incapable of living in water.

Many fresh waters, especially tropical ones, are temporary in nature and organisms in them can withstand desiccation for long periods by encysting (e.g. protozoans, tardigrades and rotifers) or forming resistant eggs (e.g. crustaceans). Other animals migrate when desiccation threatens (insects, amphibians); others bury themselves deep in the the substrate in

Figure 6.20   One of the fairy shrimps *Branchipus*.

a torpid state (e.g. lungfish) or secrete a protective layer of mucus around them.

The fairy shrimp, *Chirocephalus*, is a characteristic inhabitant of temporary pools in many parts of the world. The eggs have resistant shells and, if the pool where they are laid dries up, they survive for long periods—years if necessary—in dry mud. When the pool fills again, the eggs hatch into larvae well adapted for feeding on the dense populations of algae which rapidly develop. The animal swims mostly on its back by means of seven pairs of foliaceous limbs, which also create a feeding current along the ventral surface (Figure 6.20). Food particles are filtered from the water by means of fine setae on the limbs, and passed along a ventral groove into the mouth. *Chirocephalus* has a rapid growth and reproductive rate, dense populations quickly building up in temporary pools which are free, initially at least, from most types of predator or competitor.

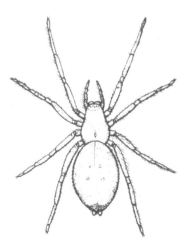

Figure 6.21   An adult of *Argyroneta*, the water spider.

The water spider, *Argyroneta*, has become almost completely aquatic, and is one of the few spiders to do so. The adaptation is mainly behavioural, for there are few morphological differences from terrestrial spiders. *Argyroneta* climbs under water using aquatic macrophytes, and among these builds a bell-shaped shelter of silk from spinnerets at the tip of the abdomen. This shelter is gradually filled and extended by a bubble of air which the animal brings down from the surface attached to hairs on its abdomen, a small bubble at a time. On completion of this shelter, the spider spends most of its time inside it, breathing oxygen which diffuses into the bubble from the surrounding water (Figure 6.21). It feeds on insects trapped at the water surface which are seized and carried down to the shelter to be eaten. Eggs are laid inside special compartments divided off from the main chamber.

CHAPTER SEVEN

COMMUNITIES AND ENERGY FLOW

A biological community may be considered as an integrated assemblage of plants and animals living together in a particular environment. The structure and composition of some communities are now well known, for example the communities of a stony stream and a weedy pond are distinct ecologically and would be categorised as such by most ecologists. However, though it is possible to draw up schemes of community classification at different levels, many of them run into the difficulty that many fresh waters are transitional in type. Elaborate classifications are fraught with difficulties, and the divisions within them are often those of greatest convenience to the classifier, and not necessarily of ecological significance.

The variation in approach to communities in the past has led to several differing concepts. The idea that the species is the basic unit in ecology (Macan, 1963) seems a fundamental one. True communities are built up of these units associated in a characteristic way, but it is essential to remember that communities are dynamic systems, maintained by the ecological interaction and flow of energy among the species concerned. The present chapter considers firstly the major communities found in fresh waters, and secondly the dynamic nature of the interrelationships within them.

The communities discussed are grouped into three types: those of standing waters, of running waters and of specialised systems. Within these the major features are outlined and discussed. The classification adopted is a broad one and intermediate categories occur. Macan (1963) pointed out that 'the time is not yet ripe for a comprehensive book on the communities of fresh water'. This is still so, and will remain the case until much more is known about the causal factors involved in the distribution and abundance of plants and animals and the processes occurring in aquatic systems.

7.1 Standing waters

7.1.1 Temporary ponds

Their small size and large surface to volume ratio means that ponds are more subject to climate than most fresh waters. Those affected most are the ones which dry up each year and develop a characteristic temporary community. Apart from desiccation, such waters are liable to large fluctuations in temperature and in chemistry (which varies according to the amount of rainfall or evaporation). There is regular mixing, high turbidity (due to disturbance of bottom deposits by wind and animals wading in to drink) and plenty of light for plant growth.

The algae in temporary pools are poor in species but rich in numbers. They include *Euglena, Phacus* and *Trachelomonas*, and in pools rich in organic matter these can occur in almost pure culture and in sufficient numbers to colour the water. Many macrophytes are adapted to temporary ponds; in the tropics *Scirpus grossus* and other species pass the dry season as tubers, while in other areas forms such as *Najas, Subularia* and *Trapa* survive as seeds. Even plants thought of as purely aquatic, like *Nymphaea* and some *Potamogeton* species, can produce terrestrial forms if the water dries out. *Montia fontana* is annual under terrestrial conditions, but biennial or perennial if aquatic conditions prevail. Versatile plants of these types are the dominant macrophytes in temporary pond communities.

Invertebrates in temporary ponds are characterised by various protozoans, rotifers, crustaceans and insects. These overcome desiccation by forming resistant cysts or eggs, burrowing deeply into the substrate, or by being very mobile as adults and so able to colonise or abandon ponds. All have fast rates of development and reproduction, building up dense populations in a short time. As with algae, invertebrate populations tend to be poor in species, but rich in numbers.

Few vertebrates occur permanently in temporary pools. Nevertheless, several amphibians thrive in those holding water long enough to allow the young to reach metamorphosis; under such conditions these animals are numerous, for the young have an abundant food supply and a lack of predators or competitors. Only a few fish can survive in temporary waters; among these are lungfish (e.g. *Neoceratodus*) and the air-breathing *Saccobranchus*. A few species of killifish (Cyprinodontiformes) are remarkable in laying eggs capable of withstanding desiccation, which hatch successfully when aquatic conditions prevail again. Various birds, especially waders and ducks, are common in temporary ponds where they exploit a rich, easily available, food supply.

7.1.2   *Permanent ponds*

Several attempts to make a precise distinction between ponds and lakes have failed, largely because there is a complete transitional series from the smallest pool to the largest lake. The very nature of seral succession precludes an absolute distinction between the two. Nevertheless, like the terms eutrophic and oligotrophic, the concept of the ponds and lakes has value when used generally, and different communities are associated with each. In general, a pond is a standing water shallow enough for light to reach the bottom all over, which never stratifies and which is too small for wind-generated waves to erode its shores. A lake, on the other hand, has profundal depths where there is insufficient light for photosynthesis, thermal stratification is common, and the shore is eroded by wind-generated waves (Figure 7.1).

One of the factors determining pond communities is the quality and quantity of plant growth there. Both algae and higher plants are normally abundant, the species present being related to nutrient status. In acid bog pools, for instance, algae are dominated by Desmidiaceae and Mesotaeniaceae, and a few characteristic genera from other groups (e.g. *Stigonema*, *Chlorobotrys* and *Zygogonium*). Certain emergent macrophytes (*Carex*) and some submerged ones (e.g. *Sphagnum* and *Utricularia*) may be the only higher plants. In more basic ponds the communities are richer, and include a wide variety of algae (planktonic, benthic and epiphytic)

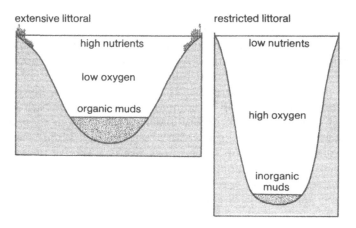

Figure 7.1   A schematic representation of the main differences between eutrophic (left) and oligotrophic (right) lakes.

and higher plants (submerged, floating and emergent). In pools which are very high in nutrients the flora is less varied and dominated by blue–green algae, or certain floating macrophytes (*Lemna, Azolla* and *Eichhornia*).

Most ponds have a rich fauna—especially ponds in the medium to high nutrient categories—characterised by protozoans (notably testaceous Rhizopoda), small crustaceans (Cladocera, Copepoda, Ostracoda), insects (Trichoptera, Odonata, Ephemeroptera, Coleoptera, Hemiptera and Diptera), molluscs and amphibians, the latter often the principal large predators. As in the benthos of other systems, nutrient-poor acid waters tend to be dominated by insects, nutrient-rich alkaline systems by crustaceans and molluscs. Some permanent ponds have fish populations and these are a major influence on the rest of the community.

7.1.3 *Lake littoral*

Though one of the most important communities occurring there (Figure 7.2), the littoral systems of lakes are difficult to define precisely, because of the diversity of conditions found—even within one lake. There is much to be said for a sub-classification within this community, but this is not within the scope of the present volume. The dominant

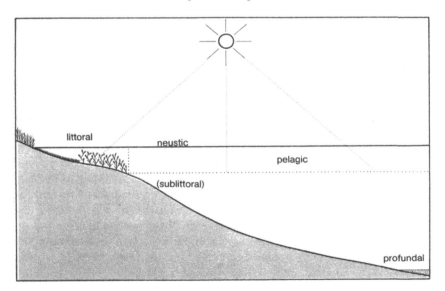

Figure 7.2 The major habitats found in a typical lake.

physicochemical features of the littoral regions of lakes are abundant light, fluctuating water levels and wind-generated waves with their mixing effect on dissolved materials and disturbance of bottom sediments. The latter can vary from fine organic muds through silts, sand and gravels to stones, boulders and bare rock. The sediments are often graded—usually from the coarsest at the edge to fine particles in deeper water.

With an abundance of light, the littoral community is the richest part of a lake for plant life—both algae and macrophytes. As well as the phytoplankton (essentially the same as in open water), there are normally large numbers of attached algae, especially diatoms, blue–greens and greens. These algal communities are very varied according to local circumstances; in shallow exposed conditions encrusting or attached forms are dominant, but where there is shelter large masses of filamentous species may develop. Many are epiphytic on other littoral algae or on macrophytes. Algal species also vary according to water depth; deeper water forms tend to be coloured red rather than green.

Macrophytes in a littoral community can range from nil in unstable sand or bare rocks to thick stands of emergent, floating or submerged species growing as deeply as light intensity will allow. On exposed shores, though emergent forms may be absent in shallow water, submerged forms will grow successfully in quieter deeper water offshore. Maximum colonisation by higher plants occurs in substrates which are stable and have a fine texture (Spence, 1967); since light, wave action and substrate all vary with depth, no one factor can be singled out as the prime control on macrophyte communities.

Invertebrates in littoral communities are as varied as the algae and macrophytes; they range from associations characteristic of truly lentic conditions to those of the nearly lotic habitats found on exposed lake shores. Typical of the former are free-swimming forms (e.g. *Notonecta*, *Corixa* and various beetles) and species associated with macrophytes (e.g. plant-boring *Cricotopus* and *Donacia*) and aufwuchs (various snails and mayfly larvae). On exposed stony shores, communities similar to those of running waters occur (though those dependent on a constant current for food, e.g. *Simulium*, are absent), while on areas of bare sand most invertebrates burrow (e.g. larval chironomids, oligochaetes and nematodes).

Fish are important in littoral communities, often determining the numbers of invertebrates there and sometimes the species too. Most lake species occur in this habitat, e.g. various Cyprinidae, Gasterosteidae, Centrarchidae and other bottom-dwelling and weed-loving species. In the littoral area, too, many larger fish, characteristic of the nekton, spawn at different times of the year, and their fry may occur here for the first part of their lives, utilising the littoral zone as a nursery area.

7.1.4 *Lake profundal*

The characteristic physicochemical conditions dominating lake profundal areas are the absence of light, the long periods of uniform temperature, the fine sediments, the absence of higher plants and any form of cover, the amount of food available as plankton or sedimenting material, and the levels of oxygen during stratification.

The lower layers of mud are normally anaerobic, the only organisms found here being bacteria such as *Methanosarcina methanica*. In the softer upper layers there is a variety of anaerobic bacteria and many fungi and protozoans. Higher plants are absent, but some colourless algae may occur, and there are often phytoplankton species, either sedimenting out during calm weather (especially heavier forms) or slowly circulating via return currents to the surface again.

Recent studies of the hypolimnetic Protozoa in lakes by Laybourn–Parry *et al.* (1989a, b) have revealed some extremely interesting facets of this part of the community. Two distinct planktonic communities were found in the eutrophic lake studied: an obligate planktonic community (dominated by ciliates and flagellates) restricted to the epilimnion during stratification, and a hypolimnetic community (Figure 7.3) of benthic migrants (e.g. *Spirostomum* and *Loxodes*). The two communities do not overlap to any extent. Some of the reasons for this have been investigated. Oxygen is toxic to *Loxodes* (Finlay *et al.*, 1986) and its effects are exacerbated by light; in addition, *Loxodes* has the ability to respire nitrogen (Finlay, 1985). Its distribution pattern has been explained as a compromise between avoiding high oxygen tensions and meeting the need for aerobic metabolism in a zone where predation from zooplankton will be minimal. It is probable that the development of two distinct communities of planktonic Protozoa is characteristic of eutrophic waters which stratify regularly. In oligotrophic and mesotrophic lakes, where the hypolimnion retains its oxygen, true planktonic ciliates are likely to occur throughout the water column.

The macrofauna is dominated by burrowing forms, notably tubificids, chironomids and sphaeriids. In addition to these, some other forms are important in certain lakes notably the larvae of *Chaoborus*, and various turbellarians, copepods, ostracods and hydracarines are present in small numbers. The profundal communities of oligotrophic lakes tend to be

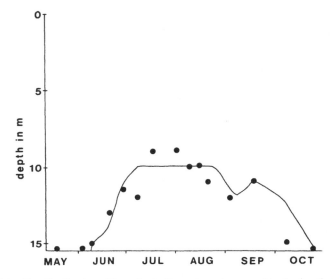

Figure 7.3   The distribution of *Loxodes* in the water column and in the benthos at 15.5 m in a eutrophic lake. The line indicates the 0% oxygen isopleth; the circles indicate the depth of maximum abundance of *Loxodes* at each sampling time (after Laybourn–Parry *et al.*, 1989b).

richer in species but poorer in numbers than those of eutrophic lakes. The classification of lakes by Thienemann (1925) points out that the profundal regions of oligotrophic lakes are dominated by *Tanytarsus*, the eutrophic ones by *Chironomus*. Most members of the community depend on mud for shelter and the organic particles in it or filtered from the water for food; a few groups, such as tanypod midge larvae, are carnivorous.

The only vertebrates in profundal regions are fish, many of which spend part of each season or of each day there. Various *Coregonus* and *Salvelinus*, for instance, inhabit the bottom at considerable depths; though sometimes eating benthos, these fish are usually plankton feeders and move up into the open water to feed at certain times, often just at dusk or dawn.

7.1.5   *Lake plankton*

By definition, plankton is dissociated from any substrate, and subject to the movement of water masses in the lake. The main physicochemical characteristics of its habitat are the varying levels of light intensity from the surface downwards, the periodic turbulence caused by wind at the water surface and, often, the periodic or permanent division of the water mass by the thermocline, with the consequent physicochemical differences between the hypolimnion and epilimnion.

Bacteria are the smallest members of the plankton, and their numbers range from less than 100 000 per ml in an oligotrophic lake (Lake Baikal) to more than 2 000 000 per ml in a eutrophic system (Lake Beloye). Though no higher plants are found in the plankton, algae are represented by a wide variety of species, especially single-celled forms (often flagellated) or small colonial species. Euglenophyceae, Cryptophyceae, Dinophyceae, Chlorophyceae, Xanthophyceae, Chrysophyceae, Bacillariophyceae and Cyanophyceae are all common. Dystrophic and oligotrophic waters tend to have high proportions of Desmidiaceae (Chlorophyceae) while many eutrophic waters are characterised by blooms of Cyanophyceae.

The zooplankton is made up of a large number of species, the dominant groups being protozoans (notably Rhizopoda, Heliozoa and Ciliata), rotifers and crustaceans (especially Cladocera and Copepoda). A few other groups, e.g. coelenterates (*Craspedacusta*), insects (*Chaoborus*) and hydracarines may be important in certain lakes. The feeding relationships within the plankton can be relatively self-contained, with zooplankton feeding on bacteria and algae, or, in carnivorous forms, on each other. As with the profundal benthos, it is true both for phytoplankton and zooplankton that oligotrophic lakes tend to be rich qualitatively but poor quantitatively as far as species are concerned; the reverse is the case in eutrophic waters.

Associated with the plankton is the nekton, made up largely of fish species capable of active movement through the water, in contrast to the passive motion of phytoplankton and zooplankton. In terms of energy flow, these fish (e.g. *Coregonus*, *Salvelinus*, *Tilapia*) are essentially a component of the plankton community, being largely dependent on its members for food.

7.2 Running waters

7.2.1 Temporary streams

Some streams periodically (often seasonally) virtually disappear. Organisms in them have to contend with the same problems as in desiccated standing waters. Various bacteria are able to withstand such drought conditions, while algae such as *Tribonema*, *Stigeoclonium* and *Ulothrix*, common when water is present, withstand drying out by forming cysts. Fully aquatic hydrophytes are rare in temporary streams, but some *Ranunculus* and *Callitriche* may continue growing as terrestrial forms; alternatively they may withstand the drought period as seeds. Hynes (1958) found that a surprising number of stream animals (including platyhelminthes, oligochaetes, ostracods, copepods, hydracarines and various insects) survived in the substrate in a stream which was dry for over two months. Other animals (e.g. mayfly, stonefly and caddisfly larva) were eliminated during this period. Many of the animals surviving drought periods in streams are small species able to exist underground as part of the interstitial fauna. Fish, other than those migrating during the wet periods, are absent from temporary streams.

7.2.2 Eroding waters

An eroding water is a running system where the current is consistent in one direction, and strong enough to erode the bottom, so that it is stony or rocky. The width may vary, but depth is normally less than 1 m. In depositing waters, on the other hand, the current is slow enough to allow

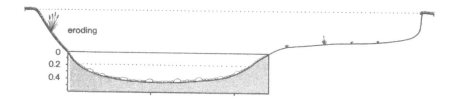

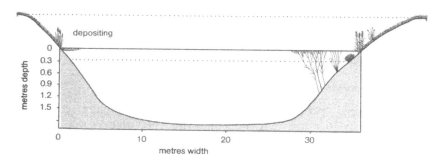

Figure 7.4 Transverse sections of an eroding (above) and a depositing (below) river.

deposition of sediment, and the depth is often greater than 1 m (Figure 7.4). Apart from current, the main physicochemical features of eroding waters are high light intensity, a clean stony or rock bottom, and the constant turbulent mixing of the water with an absence of chemical or thermal gradients. The main source of primary production is often not within the stream itself, but elsewhere (i.e. allochthonous material in the form of leaves and other plant remains). The biological communities associated with eroding and depositing waters are quite distinct (Figure 7.5).

Few free-floating algae occur in eroding waters and the community is dominated by benthic encrusting (e.g. *Rivularia*) or streaming (*Cladophora*) species. Macrophytes are relatively few in nutrient-poor streams and often restricted to a few lichens or bryophytes (e.g. *Fontinalis*), but in nutrient-rich alkaline steams, where the current is not too strong, extensive growths of some species (e.g. *Myriophyllum, Callitriche* and *Ranunculus*)

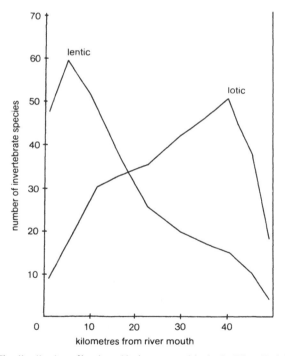

Figure 7.5   The distribution of lentic and lotic communities in the River Endrick, Scotland.

occur. Characteristically, the forms involved are all narrow-leaved and completely submerged. Broad-leaved plants, especially those with floating or submerged leaves, are atypical of eroding streams. Butcher (1933) classified the plant communities of running waters in Great Britain into five classes:

(1) Torrential, mainly on rocks, e.g. *Fontinalis* and *Eurhynchium*.
(2) Non-silted, mainly on stones, e.g. *Ranunculus* and *Myriophyllum*.
(3) Partly silted, mainly on gravel, e.g. *Potamogeton* and *Sagittaria*.
(4) Silted, mainly on silt, e.g. *Potamogeton* and *Elodea*.
(5) Littoral, mainly on mud, e.g. *Glyceria* and *Sparganium*.

Only the first two of these are eroding water communities, the others being more characteristic of depositing systems.

Invertebrates in eroding streams are typified by plecopterans, certain dipterans (e.g. Simuliidae) and trichopterans (e.g. Hydropsychidae); there are many genera, too, of ephemeropterans, chironomids and benthic crustaceans which are common. Fish are important, especially small species (Cyprinidae, Cobitidae), or the young of large ones, such as Salmonidae, which use eroding streams as nursery grounds for their eggs and young.

7.2.3 *Depositing waters—littoral*

Depositing waters are characterised by slow unidirectional current, often carrying material in suspension which is deposited where the current slows. In contrast to eroding systems, depositing waters can be divided into two distinct communities, essentially equivalent to the littoral and open-water regions of lakes.

The littoral zone of depositing waters occurs as a well-defined strip on either side of the river and is dominated by macrophytes, especially emergent and floating-leaved forms. The physicochemical conditions characterising this habitat are a soft silty bottom and water shallow enough to allow the penetration of light for rooted plants' growth. The current is slow and, though unidirectional, may be contrary to that in the main channel if the system is a backwater. The community is similar to that found in ponds or quiet lake littoral zones.

The algae in the littoral region of depositing waters are often filamentous forms, mainly benthic or epiphytic on the macrophytes which include submerged (*Potamogeton, Elodea*), floating-leaved (*Nuphar, Lemna*) and emergent (*Sparganium, Typha*) species. In acid waters, *Juncus* and *Carex* are often dominant, while *Nasturtium* or *Apium* is characteristic of hard waters. Invertebrates in this community are many and varied, including most important freshwater groups. Among the vertebrates, a wide variety of fish is found, including many small species and the young of larger species, which, though themselves living in deeper water, spawn in the plants here, which provide shelter for the eggs and young. Some amphibians and reptiles (mainly in the tropics), and numerous birds, are common members of the community.

7.2.4   Depositing waters—open

In an open depositing river channel, conditions are governed by the current and the particulate material carried from upsteam. Much of this deposits on the river bed, forming the major substrate for plants and animals. Conditions may be unstable during spates, when much of the deposits may shift downstream. The organic fraction of the particulate matter carried by the water is a major source of food for the animal community, while the plant community is determined by the substrate and the amount of light which manages to pass through the turbid water to reach it. Chemical conditions are usually rich, though in certain cases there may be a shortage of oxygen.

Apart from plankton species carried along in the water, the main algae found are benthic, usually filamentous, forms. Where sufficient light is available, macrophytes characteristic of silty substrates occur (*Callitriche* and *Potamogeton*). Emergent and floating-leaved forms are normally absent.

The invertebrates here, especially where higher plants are scarce, are similar to those dominating the profundal regions of lakes, i.e. mainly burrowing forms. Thus, various tubificids, bivalve molluscs, larval chironomids and some other burrowing insects (e.g. *Sialis*) are common. The organic material in suspension and sedimenting is food to many of these animals, and dense populations of filter-feeding species (e.g. *Pisidium, Unio* and *Anodonta*) may develop (Figure 7.6). The fish fauna is varied (but has few Salmonidae or other lotic forms) and Cyprinidae and other bottom feeders, such as suckers and catfishes, are common. Various predators (e.g. *Esox*) occur here too.

Figure 7.6   The freshwater mussel, *Anodonta*. Annual growth rings are clearly visible; from such features and data on weight and other measurements, energy budgets can be calculated.

7.3   Specialised communities

7.3.1   *Neuston*

Neuston is the assemblage of plants and animals found at the water surface, both above and below the air/water interface. Ruttner (1963) restricts his definition of this community to small organisms of microscopic dimensions, but most authors include all plants and animals associated with the water surface. Characteristics of this environment are an abundance of light (except where there is shading by overhanging vegetation) and constant availability of oxygen. Organisms of the neuston are vulnerable to surface currents, wind and wave action; consequently they are not found in fast-flowing waters or exposed lakes. Many of the animals are dependent for food on terrestrial material blown on to the water surface.

In highly organic waters, bacteria may occur in such numbers that they form a scum on the surface. Often associated with them are various algae (e.g. *Chromatophyton* and *Navicula*) and protozoans. Many higher plants are successful, and forms such as *Azolla* and *Lemna* have been discussed, as have several of the surface invertebrates (e.g. *Gerris* and *Gyrinus*). Some flatworms and snails occur as temporary members of the neuston. Though not often considered as such, many aquatic birds (e.g. ducks and grebes) are just as much members of the neuston as small organisms such as *Gyrinus*.

7.3.2 *Subterranean systems*

Subterranean systems are much more common than previously believed. The waters concerned range from the narrow interstitial spaces found among waterlogged sands and gravels (organisms found here are known collectively as psammon) to large waters, both standing and running, found in subterranean cave systems. The environmental parameters dominating such systems are the absence of light, the relative constancy of water temperatures and the low levels of primary food available.

Where there are extensive systems of subterranean fresh waters, a variety of highly adapted organisms occur, mainly animals because of the absence of light, though some fungi occur and provide food for animals. The animals are characteristically blind, but sensory tactile structures are well developed. Animals are pale or transparent in colour, and usually rather primitive compared with forms living under more normal conditions. Paucity of food usually means that populations, especially of larger animals, are small.

A variety of protozoans is found in subterranean communities. Among other invertebrates, the crustaceans *Bathynella* and *Niphargus* may be common, while various blind fish (e.g. *Anoptichthys*) and amphibians (e.g. *Proteus*) are found only in such conditions.

7.3.3 *Thermal systems*

Heated waters from industrial effluents are an increasing factor in human influence on natural waters in many parts of the world. There are, however, natural hot waters in many places with temperatures ranging from 30 to 60°C or even higher. The only organisms occurring in such waters are a few specialists which have managed to adapt to such high temperatures, for few normal freshwater organisms can tolerate temperatures of over 40°C. A variety of lower organisms are found, but not many higher ones.

Some bacteria can exist at temperatures greater than 75°C in thermal waters. Though most groups of plants have representatives which can live at temperatures of 30–35°C, only algae can live at temperatures higher than this—the most common among these being blue–green forms. In a series of waters in a thermal area studied by Geitler and Ruttner (1935), fourteen species of Cyanophyceae were found above 45°C, ten above 50°C, seven above 55°C, but only three above 60°C. Copeland (1936) gives a list of the tolerance limits of various algal and other groups, including several Myxophyceae reputed to be able to live at temperatures over 80°C.

Only a few invertebrates (mainly various protozoans, rotifers, nematodes, annelids and insects) can tolerate temperatures above 40°C; Johannsen (1932) found larvae of the midge *Dasyhelea tersa* in Java at temperatures just over 50°C. Few fish exist in true thermal waters, though some killifishes (*Cyprinodon*) and a few Cyprinidae live in waters above 35°C.

7.3.4 *Saline waters*

Small standing waters of high salt content are common in hot dry areas in various parts of the world, especially where there are inland drainage basins and the surrounding soils are rich in salts. Such waters are highly alkaline, containing large quantities of calcium, sodium and potassium, carbonates, chlorides and sulphates. The organisms found in such lakes, especially where the concentration of salts is above 10%, are restricted and consist mainly of a few algae (largely Chlorophyceae, though some Myxophyceae, Bacillariophyceae and Euglenophyceae occur too), protozoans and specialised forms of Diptera (e.g. *Ephydra*) and Anostraca (e.g. *Artemia*).

The water in most fresh waters usually reaches the sea eventually; the point at which fresh and salt waters meet is often the expanded mouth of a river, commonly called an estuary. In estuaries there can be colossal variations in conditions, especially salinity (which can vary from fresh to sea water), turbidity, temperature (fresh water is normally warmer than the sea in summer, while the reverse is the case in winter), water level, and current (both velocity and direction may vary tidally).

Compared with many other waters, relatively fewer species occur in estuaries; however, those which tolerate the extreme conditions there are often found in large numbers. There are many algae, including various Chlorophyceae, Diatomaceae, Myxophyceae and Euglenophyceae, but few higher plants tolerate estuarine conditions. In the north temperate zone the most characteristic of these are species of *Najas*, *Scirpus* and *Zannichellia*.

The main substrate in estuaries tends to be a depositing one of sand and silt, consequently invertebrates (as in similar substrates in fresh waters) tend to be dominated by burrowing forms. Apart from various protozoans, nematodes may be abundant, and annelids, crustaceans (e.g. *Corophium*) and molluscs (e.g. *Hydrobia*) may also occur. Typical fish in estuaries are the diadromous species passing from fresh water to the sea or vice versa; less transient members of the community include various Clupeidae (herring and shad), Cyprinodontiformes (killifishes) and Mugilidae (mullets).

7.4 Community structure

Elton (1927) was one of the first ecologists to emphasise the dynamic nature of communities, but showing at the same time that they all adhered to certain broad principles. The basic ideas concerning interrelationships within ecosystems have evolved from elementary ideas about food chains, through food webs and energy flow pathways, to sophisticated computer simulation models of such systems. The contribution of work from fresh waters to this field has been considerable, for the communities involved are often less complex and more self-contained than those of marine or terrestrial ecosystems.

Certain concepts involved in energy-flow systems must be defined clearly if confusion is to be avoided; among the most important of these are the terms standing crop and production. In many older publications the latter term is used where only the former is intended. The standing crop of a population of plants or animals is a measure at one point in time, or the average of measurements at different points in time, of the numbers or quantity of organisms present; it is usually expressed in terms of numbers or weight per unit area, and is essentially a static concept. True production, on the other hand, is a more dynamic idea, for it represents a measure of the amount of material or energy metabolised or stored by an organism or population of organisms within a period of time. It is usually expressed in terms of weight of material or units of energy per unit area per unit time. Gross production is the total amount of chemical energy metabolised; net production is the actual amount which goes into growth. The difference (usually the greater part) is used up in metabolism, normally assessed by measuring the respiration of the species concerned.

7.4.1 Trophic levels

All integrated ecosystems have certain general features and levels of organisation in common. The ability to support quantities of life of various kinds depends initially on the amount of energy available, linked closely with the trophic nature of the system (Figure 7.7). Organisms within the community transfer this energy from one grade to another, through what are known as trophic levels. There are three main classes of organism

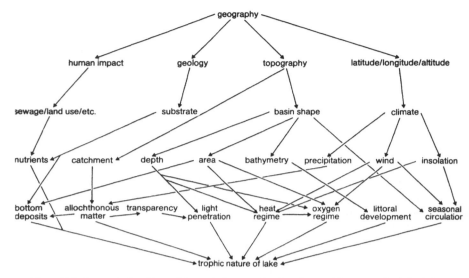

Figure 7.7   Interrelationships of the main factors affecting the trophic status of a lake (after Rawson, 1939).

concerned with this transfer of energy:

(a) *Primary producers*, which utilise the energy from solar radiation and available inorganic nutrients to produce more energy-containing (plant) material. The main organisms involved here are those capable of photosynthesis—certain bacteria and most algae and higher plants.

(b) *Consumers*, which cannot synthesise material from inorganic sources only, but have to rely on organic substances already elaborated by primary producers. There are two main types of consumer: herbivores (secondary producers which feed only on plant material) and carnivores (which feed only on animal material). In practice the situation is often complex, normally involving additionally omnivores, which feed on both plant and animal material, and various parasites which may utilise organisms at all levels (Figure 7.8).

(c) *Decomposers*, which attack other organisms (usually after they are dead), breaking them down into simpler compounds and releasing many of the inorganic salts again, making them available to primary producers. The main organisms involved in the decomposer process are heterotrophic bacteria and fungi.

7.4.2 Food chains and food webs

Simple systems, formerly known as food chains, are rarely found in isolation in nature. In a common ecosystem such as a stream, for example, a basic part of the energy flow pattern might be through a benthic alga (utilising available solar radiation and nutrients) which is eaten by a caddis larva, which in turn is eaten by a fish such as a minnow. This simple chain is, however, complicated by a variety of cross-links; it is likely that invertebrates other than caddis larvae eat the alga, and that fish other than minnows eat the caddis larva. Since too, many animals are opportunists as far as feeding is concerned, and rarely restricted to one type of food, it is likely that the caddis larva will browse on algae other than the species under consideration, and that the fish concerned will eat various invertebrates additional to the caddis larva. Many invertebrates are omnivorous or carnivorous, and it is likely that the caddis larva may eat other invertebrates (e.g. mayfly larvae) which feed on the alga, or themselves be eaten by predacious stonefly larvae, these in turn being eaten by fish. All the organisms concerned may well be attacked by parasites of one kind or another (Figure 7.9).

Even in the most complex situation, however, it is still possible to consider principles common to the structure of the communities and the flow of energy within them. The concept of trophic levels has been considered. Linked with this is the idea of the Eltonian pyramid of numbers (Figure 7.10); this conception is dependent on the fact that in any energy-flow system the quantities involved tend to be less and less between primary producers and top consumers. Elton (1927) pointed out that 'animals at the base of a food chain are relatively abundant, while those at the end are relatively few in number, and there is a progressive decrease between the two extremes'. This simple concept may be modified in different ways, depending on the structure of the community concerned and its stability in space and time. Its apparent contradiction of the concept of Lindeman (see p. 196) lies in the inefficiency of most biological systems—for rarely is all the food eaten by an animal assimilated and available as energy; moreover a great proportion of the energy intake of animals is utilised, not for growth, but for basic metabolism. The efficiency of many systems is surprisingly low, and top consumers may have available to them only a minute fraction of the energy originally entering the system as solar radiation.

In many communities, the pyramid of numbers lives up to its name, with large numbers of primary producers, fewer secondary producers and still fewer tertiary producers. In other situations, however, there may

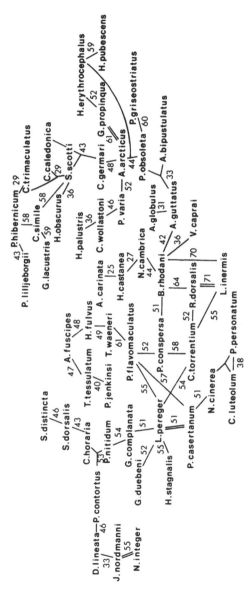

Figure 7.8 Association analysis of common freshwater invertebrates occurring in a wide range of fresh waters within one geographical area. Each species is linked to that with which it is most highly associated (Fager, 1957), the relevant values being indicated between them as percentages (from Maitland, 1979).

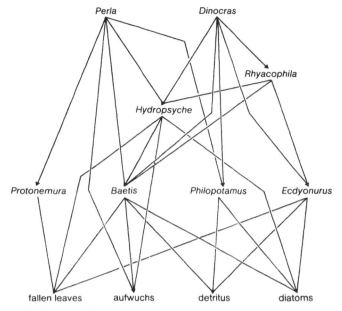

Figure 7.9   Part of a food web in a Welsh mountain stream (after Jones, 1949).

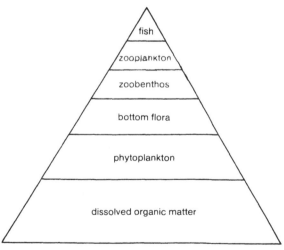

Figure 7.10   The classical pyramid of organic matter at different trophic levels in a lake. The proportional areas of the triangle approximate to the weights of the components in Weber Lake, U.S.A. (after Juday, 1942).

apparently be more secondary producers than primary ones, or, rarely, more tertiary consumers than secondary ones; such apparent paradoxes are resolved by considering the dynamic nature of the system concerned. For example, though at any one time there may be more material in the second trophic level than the first, primary production can be a rapid process, and the plants concerned can be producing large amounts of material which is constantly cropped off by the secondary consumers. It is clearly better to consider such systems as dynamic ones, with quantities of energy flowing through them, each trophic level having less available energy than the one below it. Extreme contradictions to the pyramid concept involving only numbers (e.g. the inverted pyramid found in most host–parasite systems) are therefore avoided.

7.4.3   Energy flow

A dynamic concept of a community therefore, involves the idea of the transfer of energy among its members, and from one trophic level to another. The first law of thermodynamics states that any form of energy can be transformed into another form, and that during the process the total amount of energy within the system remains the same. Lindeman (1942) has developed the idea of the trophic dynamic aspect of ecology, assuming that in systems in equilibrium the laws of thermodynamics hold for plants and animals, and that these can be grouped together into three trophic levels—primary, secondary and tertiary producers.

In early, rather crude, energy-flow budgets, decomposition data were based on the early work of Birge and Juday (1922), who analysed the amounts of material likely to be indigestible and therefore passed out with the faeces. Respiration figures were taken from the few available in the literature, while predation was estimated by calculating the energy requirements of the next trophic level. The comprehensive studies discussed below have utilised much more sophisticated techniques for obtaining these estimates, and in most cases values were obtained from the organisms in the actual ecosystem. In complex systems a complete picture of the pattern of energy flow can only be obtained from data on the population dynamics, life cycles, food, assimilation and respiration of the community collected over many years; this implies a reasonable degree of stability in the system being studied.

The element of time in the concept of production is an important one, offering a more useful method of comparing aquatic systems from the energy standpoint than one concerned only with standing crop. It is possible to have two systems with similar standing crops at any one trophic level but for the production of one to be several times that of the other. The rate of production is important both ecologically and economically, as it is only when a true understanding is available, of the effect on the system of the alteration of various parameters, that the maximum production potential can be realised. A classic example of this is found in certain fish ponds where rapid breeding causes the population to build up to such an extent that there is only just enough food available for the fish to metabolise, but practically none to allow further growth. Net production may then be very low, though the standing crop is high. There are two main ways of increasing production in a situation like this:

(a) Additional energy may be added to the system, either directly as food for fish, or indirectly by adding nutrients to increase primary production, or

(b) A portion of the fish stock may be removed, allowing the remainder additional food to assimilate and use for growth.

Both methods are widely used in the artificial production of fish and other crops by humans.

It is clear, therefore, that one of the best ways of understanding a community and the processes involved within it is to produce a total energy budget for the system concerned. However, it will be apparent that a task of this nature is extremely complicated, and one which is virtually impossible to carry out in large complex systems. It is no coincidence that the earliest and/or the most complete energy-flow studies have been concerned with relatively simple ecosystems. One of the few attempts at drawing up an energy budget prior to the hypotheses of Lindeman (1942) was that of Juday (1940), who synthesised the results of two decades of work on Lake Mendota. The community studies by Teal (1957) and Odum (1957) were relatively simple in that both were in springs with a simple community structure and almost constant temperature (Figure 7.11). A further, rather more complex, study by Teal (1962) was carried out on a salt marsh ecosystem. More complex and ambitious studies have since been carried out, notable among these being those on the River Thames (Figure 7.12), Loch Leven (Figure 7.13) and other waters as part of the International Biological Programme. This programme, in addition to giving a special impetus to the methodology necessary for quantitative studies, provided a considerable insight to the processes controlling the flow of energy and the structure of many aquatic ecosystems.

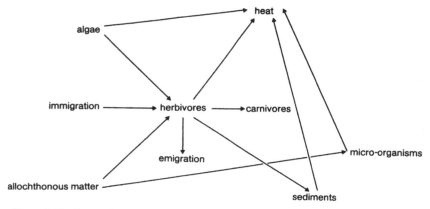

Figure 7.11   Energy flow in a temperate cold spring—Root Spring, U.S.A. (after Teal, 1957).

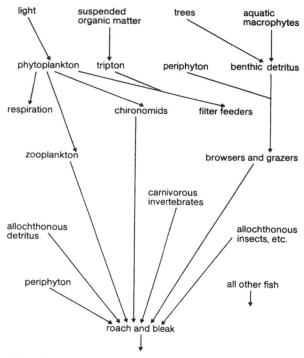

Figure 7.12   Energy flow in the River Thames, England (after Mann, 1964).

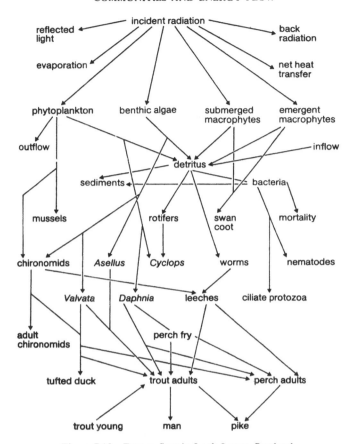

Figure 7.13 Energy flow in Loch Leven, Scotland.

As an example of one of the simple systems studied, the energy flow in the specialised community found in sewage lagoons in Oregon (Figure 7.14) may be cited (Kimerle, 1968). These systems were dominated by the larvae of the midge *Glyptotendipes barbipes* and no fish were present. All the important parameters necessary for drawing up an energy budget were monitored regularly. Though some of the energy was brought into the system via sewage entering the lagoon, the bulk was due to solar radiation. Most of this was quickly lost again through reflection, evaporation and other causes, but gross primary production utilised about 20% of the total. About half of this production was transferred to the rest

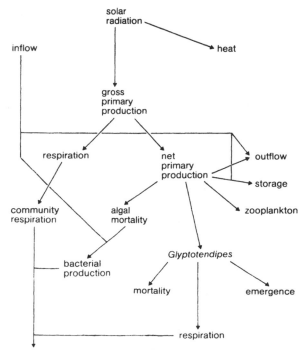

Figure 7.14   Energy flow in a sewage lagoon in Oregon, U.S.A. (after Kimerle, 1968).

of the system, most of it through death to bacterial decomposers, but a significant proportion transferred to the population of midge larvae, and lesser amounts to zooplankton, storage and loss to the lagoon system. As with all the organisms in the system, much of the energy taken in by the midges was used in respiration, but just over 33% went into net production, part of which was lost to the lagoon by the emergence of the adult midges, the remainder passing after mortality to the bacterial decomposers. The completion of the International Biological Programme has undoubtedly led to a fuller understanding of aquatic communities and the flow of energy within them.

CHAPTER EIGHT

FRESH WATER AND HUMANS

Fresh water has been used by humans from earliest times, at first only for drinking, but later for fishing and navigation. The majority of settlements in many countries are related to spring lines and other sources of pure water. With improving sanitation, water was used for cleaning and removing domestic wastes, as well as for irrigation in agriculture. Within the last two centuries, improving standards of living, increased sophistication of agricultural methods, industrial development and production of hydro-power have meant that water has become more and more important to humans. Further, extended leisure time in modern societies has increased pressure on recreational facilities for aquatic activities like angling, wildfowling, sailing, swimming, water-skiing and power-boating.

Fresh water is essential to successful modern societies, and a useful way to compare living standard of different nations is to examine the per capita consumption of water. Countries with a high standard of living (e.g. Norway, Sweden and Switzerland) show a high requirement for potable fresh water (*ca.* 100–500 litres per head per day) compared to poor countries. The ever-increasing standards of living in most countries and the increase in human population combine to make heavy demands on freshwater resources, and result in conflicting interests in available water. Production of fresh water from the sea by desalination and recycling of water after treatment by sewage works are among the approaches being adopted to combat this problem.

The increase in the quantity of water required for domestic and industrial (including agricultural) purposes, as pointed out by Rafferty in 1963, shows no sign of lessening and makes it necessary to consider the whole question of water resource and supply on a more extended basis. Water conservation means the preservation, control and development of water resources (by storage, prevention of pollution or other means) to ensure that adequate and reliable supplies are available for all purposes in the most suitable and economic way, whilst safeguarding legitimate interests. Originally, only extremely pure waters were considered as potential sources, and this emphasis on quality led to excessive exploitation of ground water and impoundments in upland areas. The increasing shortage of water, and the improving means for treatment of contaminated water, have meant that quantity and not quality is often more important; sources which would not have been considered in the past are now accepted and treated for public supply.

In arid areas, particularly those far from large natural supplies, the problem of supplying large volumes of suitable water may be expensive or insoluble. Even in areas of adequate rainfall, seasonal variation may cause difficulties, low rainfall creating a shortage, high rainfall causing flooding. This situation is typical of many countries, where the remedy is one of national expenditure on the development and conservation of water resources. Campbell (1961) has pointed out that, in Scotland, the public water supply services then provided about 1.72 million cubic metres a day for domestic and industrial needs. Needs in the foreseeable future are unlikely to be more than 2.83 million cubic metres a day; this represents only about 1.25% of Scotland's overall water resources, of which nearly half are classed as gravitational.

8.1 Water resources

The initial source of water is precipitation, and a study of the rainfall and hydrological cycle is a necessary preliminary to assessing the water supply potential of any area. Only a relatively small proportion of the total rainfall in a large geographic area is readily available for water supply; major losses occur from evaporation during precipitation, from the ground or open water and from transpiration from vegetation. Colossal amounts of water flow into the sea directly. Utilisation of this available rainfall can involve collecting it as surface water (by intakes or pumping from rivers, or by piping from suitable lakes or reservoirs) or as ground water (by utilising springs, or by sinking wells).

The most suitable areas for utilising existing lakes or establishing reservoirs are in mountainous regions where natural systems are oligotrophic. Waters in highland areas are especially suitable for domestic supplies in that there is less pollution than in lowlands. Also, the initial rainfall is higher in highland than lowland areas. Oligotrophic waters contain little suspended matter (especially algae) and require little filtration. Though the geographic regions most suitable for water supply bodies are often far from areas where water is most needed, the altitude of such systems means that water will readily pass by gravity, obviating the need for expensive pumping. As well as the chemical nature of the catchment, its physical geology (especially where man-made reservoirs are concerned) is important. Local geology affects the amount of direct run-off, the potential underground losses and the mechanics and cost of dam construction.

Ground water has fallen on the earth as rainfall and then percolated through soil and rocks to collect over impermeable strata where it may remain for long periods. The surface of this ground water within the earth (comparable to the surface of a lake) is known as the water table, and may fluctuate according to losses from evaporation and underground run-off or additions from rainfall. Other than where large springs appear at the surface, the collection of ground water involves the sinking of a well and then pumping water which drains into this from the surrounding pervious strata (known as an aquifer).

River systems, especially in lowland areas, are less desirable as sources of potable fresh water than either of the two preceding types. Normally pumping is necessary and the water often carries considerable suspended solids and may be polluted and require considerable treatment. Nevertheless in some areas rivers are becoming the most important source of supply. The intake pipes are strategically placed so that they will always be under water, but will be little affected by strong currents or silting. It is common practice to instal a weir to raise the river level for abstraction and to deflect the water (especially at low flows) into the intake. Both weirs and dams are obstructions to migratory fish, and fish ladders may need to be incorporated.

8.1.1 *Reservoirs*

To make economic use of water resources—especially in areas with low or variable precipitation—it is essential to construct adequate reservoirs. These form a component in most water supply systems, and in many areas there are now more reservoirs than natural lakes. The recent construction of very large systems like the Kariba and Aswan Dams makes these among the largest freshwater bodies in the world. For many purposes (irrigation, some industries, navigation and power production) the quality of water is not of primary importance, the main factors being the amount and head available, and the efficiency of the system in relation to catchment hydrology. With reservoirs for domestic supplies, on the other hand, it is essential to consider not only the physical and engineering aspects of the system but also the biological ones.

Within a reservoir system it is often desirable to have an initial settling reservoir to sediment incoming silt. Main reservoir basins may be shallow or deep; if shallow, there is less likelihood of stratification, but there may be a higher level of algal production. In deep reservoirs, algal production will be lower (because less water is exposed to light) but stratification is probable and may lead to anaerobic conditions in the hypolimnion. The resulting odour and taste may make water here unsuitable for immediate domestic supply. It is likely to be richer in nutrients than water in the epilimnion, and if compensation or spill water from the system can be drawn off below the thermocline the nutrient level of the reservoir will be reduced and algal problems lessened.

8.1.2   Flood protection

Most human problems connected with fresh water concern shortages in quantity or quality. With flooding, however, excess is the problem, and one which causes tremendous damage to life and property in parts of the world; in a number of areas, massive and expensive flood prevention schemes have been installed. Many problems connected with flooding arise from the fact that the most desirable industrial, agricultural and residential areas lie on flood plains close to rivers, and that with the increase in river canalisation and land drainage schemes, water falling on uplands can find its way to the flood plain very quickly.

There are several ways of preventing flooding in lowland areas (Figure 8.1). River training works (e.g. stone or concrete walls, willow piling or groynes) can be constructed to give permanency to the river channel and prevent damage by erosion. The area flooded may be reduced by constructing flood defences (earth embankments or concrete walls). The water level in the main river can be reduced by providing flood relief channels, by enlarging the main channel (either with embankments or by dredging and widening), or by intercepting the flood water before it reaches the danger area, and diverting it through a new intercepting channel or into a flood storage reservoir. In some cases pumps are used to deal with urban flood problems. Many modern schemes incorporate several of these methods. Careful afforestation in upper catchment areas is one long-term method which has been tried in several parts of the world, while multipurpose reservoirs serving both flood control and water supply are becoming common.

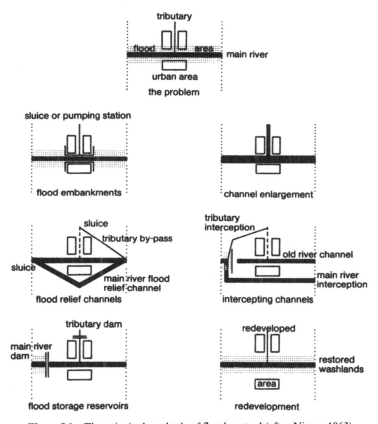

Figure 8.1 The principal methods of flood control (after Nixon, 1963).

8.1.3 Man-made systems

Throughout the world, humans have manipulated many natural freshwater systems for water supply, transport and other purposes. Though there are few completely artificial running waters (other than canals—if these can be categorised as such) there are many thousands of standing waters of all sizes from very small fish ponds to gigantic reservoirs such as the enormous new lakes in Africa. In some cases these new waters have become fairly natural and integrated well with the local ecology; in others there have been major ecological and environmental problems.

Where a barrier is put across a river in the form of a dam, ecological conditions change substantially both upstream and downstream of the obstruction. Upstream, conditions bear little relation to those pertaining before the impoundment and the aquatic system changes from a lotic to a lentic one. There are often enormous post-impoundment changes in the populations of plants and animals in the new reservoir which take years to stabilise. Downstream, the river loses much of its dynamic nature, with reduced or controlled flows and substantial changes in physical and chemical conditions.

In Africa particularly, some enormous artificial lakes have been created in recent times (Obeng, 1981). Between 1958 and 1968, four major rivers were dammed to create new lakes: Lake Kariba on the River Zambezi (1958), Lake Volta on the Volta River (1964), Lake Kainji on the River Niger (1968) and Lake Nasser on the River Nile (1968). Only now, after several decades is it becoming possible to assess the ecological, economic and sociological advantages and disadvantages of these enormous artificial waters.

The development of such large reservoirs has followed a characteristic sequence in most parts of the world. The decisions to build them are political and economic. During construction, local communities are dislocated and resettled elsewhere. As the reservoir fills their former settlements, and farmlands, surrounding areas of forest and wildlife are also flooded. Once the reservoir is full there are enormous changes in its ecology which take years, often decades, to stabilise. Many species of plants and animals, formerly abundant in the river system, disappear. Others flourish, and indeed some (including new species which have been introduced intentionally or accidentally) explode, creating enormous population densities in a very short period of time.

Some of these changes are advantageous in human terms, for instance there is often a boom in fish production: in Lake Nasser, the fish catch rose from 1550 tonnes in 1966 to 4545 tonnes in 1969. However, where humans have deliberately introduced alien species into new systems, the results can be disastrous. The water hyacinth *Eichhornia crassipes* and the water fern *Salvinia auriculata*, both floating plants from South America, have been introduced to waters in Africa, Australia and other tropical regions where they have become major pests on canals and reservoirs, forming enormous floating masses which impede navigation, hinder fishing and substantially affect water quality. At one point in Lake Kariba, the water fern covered over 21% of the surface of the lake. The water hyacinth is now a declared noxious weed in Australia.

Eventually, these systems settle down ecologically and most of the enormous perturbations of the early years disappear as the communities in the new lakes stabilise and gradually become more like natural local systems. In the rivers below the dams important changes take place, especially related to flow regulation and the carriage of suspended solids. As in all natural lakes, these new reservoirs act as sumps for suspended solids and so most of the silt carried into them—sometimes in enormous amounts during floods—settles out there and starts to fill in the new basin. The main functions of these new reservoirs are usually to store water (perhaps to be transferred elsewhere) and to produce hydro-electricity, both of which mean regulation of the flows downstream. This regulation, together with the loss of silt has changed some lower rivers and their flood plains dramatically. A good example is the lower River Nile where the alluvial flood plains are much less rich than formerly and the delta area at the mouth is changing physically and has a very much reduced fisheries production.

Thus the ecological consequences of the construction of dams on major river systems are substantial and often complex. Many of the problems are still far from being understood and there is an urgent need for further study so that the planning and management of these enormous systems can benefit from ecological advice from the conceptual to the management stage. The projects concerned can be of enormous economic importance to the areas concerned and with better ecological input to planning many of the disasters of the past could be avoided.

8.2   Water use

8.2.1   *Domestic*

Organisations supplying water for human consumption are concerned with producing as economically as possible adequate volumes which are free from objectionable odour and taste. This water must also be clear and free from harmful mineral substances, as well as disease organisms (e.g. certain bacteria and viruses). Among the important human diseases which are known to be spread by water are bilharzia, amoebic dysentery, gastro-enteritis, leptospirosis, infectious hepatitis, cholera and typhoid. Large amounts of some inorganic salts may cause dental fluorosis (excess of fluoride), methaemoglobinaemia in infants (excess of nitrate), and poisoning due to lead or other heavy metals. Adequate precautions to prevent pollution of the water supply, and efficient treatment before consumption should be sufficient to avoid troubles from most of these causes.

Many reservoirs are constructed for storage, so that water produced by high rainfall may be available for supply during dry periods. In some cases, however, especially with river waters, storage in large reservoirs is a preliminary form of treatment, when the water loses much of its suspended matter and harmful organisms are destroyed. Colour and some forms of nitrogen may also be reduced, but storage can involve an increase in the algae and other plankton present. Most waters for domestic supply require further treatment to remove suspended and dissolved solids, colour, taste and odour. This may involve sieving, coagulation, sedimentation, filtration, sterilisation and softening.

(a) *Sieving.* Water from a storage reservoir or river is subjected to coarse screening which removes all debris (twigs, leaves, etc.) which might clog up machinery. The function of such screens (which have 2–3 meshes per cm) is not to protect the consumer, but the equipment involved in further processing. Micro-strainers form the next part of the sieving process; they consist of wide areas of extremely fine-mesh material through which water is passed (Figure 8.2).

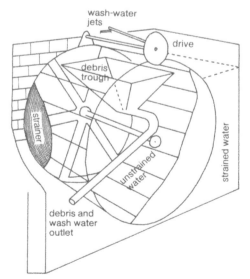

Figure 8.2 Layout of a typical micro-strainer unit for filtering domestic water supplies.

(b) *Coagulation.* If a chemical coagulant is added to water it produces a floc which aids sedimentation and provides a fine filtration film on sand and other straining beds. The most widely used coagulants are salts of aluminium and iron (e.g. aluminium sulphate, sodium aluminate and ferrous sulphate).

(c) *Sedimentation.* Like sieving, sedimentation removes large amounts of solid material from the water, thus easing the burden on subsequent treatments. It may be carried out in normal reservoirs or in special sedimentation tanks. River waters with high amounts of suspended solids lose much of these in a storage reservoir and, if this has a reasonable retention period, this alone may be sufficient to remove solids prior to filtration. In many cases, however, further sedimentation is necessary, and special continuous-flow sedimentation basins made from concrete are used.

(d) *Filtration.* Sand filters are the most common method of final removal of suspended materials before water is passed to the consumer. Where the water is of good quality initially, this may be virtually the only treatment, but where large amounts of suspended solids are present, sand filtration may be preceded by sieving, coagulation or sedimentation. Sand filters are contained within basins of concrete or brick and built up of a layer of sand (about 50 cm deep) overlying a layer of graded gravel (also about 50 cm deep) which lies on specially designed underdrains. Water brought into the system is filtered as it passes downward through the sand and gravel. Sand filters are of two kinds—slow and rapid—which operate in slightly different ways: in both cases the process involved is more complicated than a simple layer of sand acting as a filter. Shortly after it starts to function, the surface of a sand filter becomes invaded by micro-organisms which form a complex association and a fine network over and among the sand particles. An invaluable function of this network is the removal not only of suspended solids but also of many oxidisable organic materials.

(e) *Sterilisation.* Though sieves, sedimentation and filters remove a high proportion of suspended solids and many bacteria, in some cases (especially polluted waters) significant numbers of bacteria pass through and must be killed chemically. Various treatments are available, but the two most commonly used are based on chlorine and ozone. The latter is effective for virus removal, but is expensive and is normally used only in specialised situations.

To apply chlorine to small water systems, chloride of lime or sodium hypochlorite is used, but in large works chlorine itself (in liquid form) is added.

(f) *Appearance.* Unpleasant tastes, odours and colours are undesirable in domestic water supplies, though rarely harmful. They may be caused by substances in the original water, or arise during treatment (e.g. algal growths during storage, excess free chlorine). Modern water treatment ultilising efficient sieving, coagulation, sedimentation, filtration and sterilisation systems is normally adequate to deal with any tastes, odours or colours, apart from a few which require adjustments in the treatment process itself.

(g) *Softening.* Water supplies from calciferous rocks, especially ground water, often contain large quantities of calcium and magnesium bicarbonates and sulphates. These waters are not harmful to humans as drinking water, but do require much more soap during washing than soft waters, and may cause furring of boilers. It is on such economic grounds that hard water is treated. Three principal methods are used to soften water:

(1) The precipitation of calcium and magnesium as insoluble salts.

(2) The ion-exchange method where calcium and magnesium are replaced by sodium.

(3) Complete demineralisation where, by ion- and anion-exchange methods, all dissolved salts are removed. Because of the high incidence of certain heart conditions in soft-water areas, over-softening of hard waters for drinking should be avoided.

8.2.2 *Irrigation*

Of the large amounts of water used by agriculture, by far the greatest percentage is used for the growth of crops, much of this being immediately passed into the atmosphere by transpiration. Only a small percentage is used for consumption by stock, vegetable washing and implement cleaning, and this is normally available from private local systems or from the public mains supply. In some countries adequate rainfall meets the needs of outdoor crops, but in many others—including some in temperate areas— precipitation is inadequate, and supplementary water must be provided by artificial irrigation. In arid areas, agriculture is possible only if adequate water for irrigation is available, and the key to the development of many

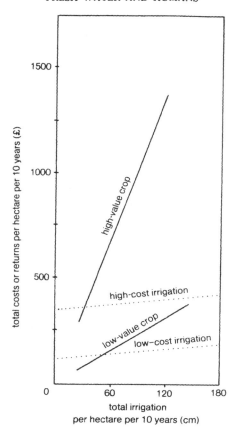

Figure 8.3 The relationship between costs (solid lines) and returns (dotted lines) in irrigating crops (after Prickett, 1963).

countries lies in providing water for this purpose at an economic cost (Figure 8.3).

Water is needed constantly by most crop plants, especially during their growing season. It is used by terrestrial plants to maintain the transpiration process by which water evaporates from the leaves and further water, together with nutrients, is drawn up from the soil. If adequate moisture is not available, the plant becomes less turgid, transpiration and water flow slow down, and growth is checked. Available soil moisture depends very much on local hydrology, and among factors influencing this are the soil types present—high values being found in loams and low values in sands. The soil moisture deficit is the amount of water needed to increase the soil moisture to its capacity; a deficit up to half the total capacity is normally permissible, but when less than this is available irrigation becomes necessary.

Irrigation water requirements are more seasonal than those for domestic water supply, and unfortunately tend to be greatest when rainfall and artificial storage reserves are lowest. Specific irrigation requirements vary according to local conditions and the type of crop grown, and must normally be calculated for each area (Prickett, 1963). Irrigation is worthwhile only where the costs of application over a period of years are covered by the increased return from the crops so grown. This is often possible by irrigating only small areas on which the intensive culture of high-value crops is practised.

Primitive irrigation techniques consist of miniaturised canal systems with numerous ditches leading off the main channels into the cultivated areas, often on a ridge-and-furrow system. Ditches and canals of this type need constant maintenance to keep them silt- and vegetation-free, and a regular head of water must be maintained to give constant flow. Originally, power to maintain this head was derived from direct labour or large domestic animals, but water pumps are now widely used. In addition to efficient pumping facilities, modern irrigation schemes utilise piping systems of distribution mains (10–15 cm diameter) and sprinklers, rainguns or spraylines leading from these.

8.2.3   *Aquaculture*

The culture of fish is a very ancient form of rearing animals for food and has been carried out in China for more than 4000 years (Lin, 1949); in many areas the methodology and systems involved have changed little over hundreds of years. In Europe, the records do not go back further than the twelfth century, since when many species of fish have been widely kept. In Great Britian, many of the early waters were stew ponds associated with monasteries, but in recent years fish farming has expanded rapidly (see below). Elsewhere in Europe, in Russia, the Middle and Far East (especially China), some parts of North America and many areas of the tropics, the rearing of fish is an important and expanding part of the economy, particularly as far as the production of protein is concerned.

Marine fish farming is now important in some parts of the world, but until recently fish farming was carried out mainly in fresh waters, or

Table 8.1 Typical fish crops from waters in different countries. Values in kg/ha.

Country	Crop
Switzerland	13
Germany	21
USA	22
Uganda	168
Java	6000

occasionally in brackish waters in some tropical areas. With efficient methods and suitable local conditions, extremely high rates of production are possible compared with other forms of protein production (Table 8.1). Moreover, fish farming is often the only practicable use for some types of ground —where, for example, drainage is poor or ground water too saline for traditional agriculture. Fish farming is often compatible with other farming: in Africa and Asia, fish and cereals are grown together in irrigation schemes, while in Europe many fish ponds are dried out in alternate years to grow cereals.

The main requirements for successful fish farming are suitable water, an adequate area for ponds (with suitable soil), local supplies of fertilisers or food, adequate labour and experience to operate the farm, and a good local market for the fish, or freezing facilities and transport to such a market. The water required depends on the fish being kept: salmonids require large amounts of cool, high-quality water, while carp will tolerate water of poorer quality and need only enough to keep the pond filled. With adequate water, a site is required where the soil will retain it or can be made to do so relatively easily. It is always advantageous if ponds can be drained, and rarely necessary to have them very deep—small ponds for young fish are rarely deeper than 1 m and very few artificial fish ponds are more than 2 m in depth (Figure 8.4).

The crops produced by farming fish can be variable and depend very much on the species of fish, the management techniques, and the amount of energy put into the system, whether in the form of solar radiation or direct feeding. All ponds have a maximum standing crop; this is not necessarily the best one for maximum production. The production potential is controlled by the maximum standing crop and the time taken by the initial stock of fish to reach this level. The natural standing crop of a fish pond may be increased several times by fertilisation or by

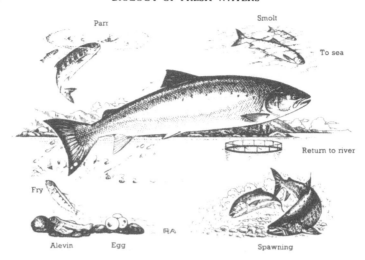

Figure 8.4 Modern fishfarming. Here, an intimate knowledge of the biology and ecology of a fish (in this case, the anadromous Atlantic salmon) has enabled it to be farmed intensively, the young stages in fresh water, the adults in the sea (from Maitland, 1989).

supplementary feeding. When the standing crop of fish approaches the maximum, the population should be harvested and the whole process started again. In Israel and some tropical countries the rates of growth are extremely fast, and it is possible to obtain high annual production by taking two or more crops per year.

It is difficult to compare the yields from different places because of the variables involved. Very high yields (1.5 million kg per ha) have been reported for carp reared in cages in running water in Japan (Hickling, 1962); such crops are usually found in specialised situations, and a comparison of more normal figures can be seen in Table 8.1. Interest in fish culture techniques is growing, and in recent years has been centred not only on traditional species but also on some of the more difficult anadromous species and several marine species (including many molluscs and crustaceans).

8.2.4 *Industry*

The location of many large industrial areas in the world depends on the availability of suitable water. The main factors concerned with such water are that it is available cheaply and is of appropriate quality and quantity. Water of high quality is required for certain industries and is often obtained from domestic water supply systems. Some processes, however, can use water of poor quality, and it can be cheaper to obtain this from boreholes, rivers, canals, private reservoirs, estuaries or even the sea, and use it untreated. Hopthrow (1963) estimated that in Great Britain about 35% of the public supply is utilised by industry, but that much more than this is obtained from other sources. There are five major industrial uses for fresh water: processing, incorporation, boilers, cooling and firefighting. In addition, all industrial water users require domestic water for employee toileting, drinking and washing.

8.2.5 Hydro-electricity

Though a form of industrial use of fresh water, the production of hydro-electricity can have such an influence within the watershed concerned that it is worth considering separately. Hydro-electric stations can be constructed only where the topography and water supply are suitable, but in mountainous areas with an adequate rainfall they make a significant contribution to power production. In 1989, for instance, though over 99% of the electricity generated in England and Wales was produced by thermal power, in Scotland (Figure 8.5) more than 30% was generated by hydro-electricity stations.

Factors favourable for hydro-electric production are a high rainfall, preferably with more in winter (when the demand for power is greatest) than in summer, sufficient altitude to give a head for generating purposes, and the absence of prolonged freezing during winter. Various hydro-electric schemes have been developed according to regional differences in topography or power economics. In some large rivers, where adequate heads of water are available, base-load stations have been constructed with a capacity to suit low river flow. Other types involve storage reservoirs (Figure 8.6); the water available to these can be augmented by more brought in by aqueducts from neighbouring catchments. Such schemes may be single-stage, with one high level reservoir, or multi-stage with a high main reservoir above a stepped system of power stations, each with its own regulating pond. Pumped-storage schemes have become important in recent years, for with the need to maintain a constant load on nuclear power stations their excess power available at off-peak periods can be used to pump water into high storage reservoirs for hydro-electric stations.

Figure 8.5   The distribution of hydro-electric power stations in Scotland (after Aitken, 1963).

8.2.6   *Navigation*

Inland water systems have always been important for human transport from earliest times, especially where local topography or vegetation made overland transport difficult. Adequate navigation channels for large vessels coming from the sea to inland ports are still important to industry and commerce in all parts of the world. Before the advent of modern air, rail and road systems, transport via inland waters was so important economically that there was substantial investment in canal waterways, mainly during the eighteenth and nineteenth centuries.

Figure 8.6 One of the problems in new reservoirs or impounded lakes is the enormous area of steep barren littoral shore created by the unnatural variable water levels (Photo: P.S. Maitland.)

Natural lakes, though useful for local navigation, are rarely of importance in national terms unless they are exceptionally large (e.g. the Great Lakes of North America) or can be connected by some form of canal system with other lakes or with the sea (e.g. the Caledonian Canal system in Scotland). The main requirements of rivers for navigation are adequate supplies of water to maintain navigable channels and to counteract shoals in estuaries. If abstraction requirements of water undertakings and industry along the river are high, storage reservoirs may be necessary. Floods are a major navigation hazard in rivers, not only because of navigational difficulties, but because the river bed is frequently scoured and deposited elsewhere in the form of shoals. Because of this and the small and shallow nature of many rivers during dry weather, large-scale navigation is restricted to the lower reaches of medium rivers, or to very large river systems (e.g. the Mississippi, Rhine and Danube) which have enormous natural channels and adequate dry-weather flows.

In many areas it has proved economic to extend the natural navigable systems, thus enabling ships to pass from sea ports right into inland industrial areas. This involves providing weirs and sluices to deepen the navigable channel and retain flood water; these barriers may also include turbines for hydro-electricity. This development and canalisation of river systems has proved successful in many countries; e.g. in North America, the St Lawrence Seaway, with only seven locks, developed a traffic of over twenty million tonnes per year within three years (Marsh, 1963).

The logical successors to river canalisation were artificial canals, often linked to natural or canalised rivers. The fundamental requirement for good canals is a supply of water to meet lockage needs, especially during droughts. Often, the size and structure of canals is determined from the outset by water availability. Apart from the basic canal channel, other channels are usually necessary to carry water to it from local springs and streams. The amount of water used in a canal system increases with traffic, as large quantities pass down the channel each time a vessel passes through a lock.

8.2.7 Recreation

The increasing human population, and the leisure time available to it, is placing an enormous demand on what is left of the natural countryside. This is especially true of fresh waters, which are often the focus for a variety of recreational activities (e.g. sailing, power-boating, water-skiing, fishing, wildfowling, bathing and general picnicking). It is unfortunate, but important to note, that for various reasons (including aesthetic ones) the most important waters for recreation are often not those already within urban areas, but those further away. This is partly because of the natural preference for clean rivers and clear (usually oligotrophic) lakes as opposed to turbid, polluted and eutrophic waters. A study of the distribution of boatyards and building developments showed that the most popular areas for commerce and tourists lie in those river valleys where the greatest number and most important nature reserves are situated (Duffey, 1962).

Many wetland areas are being actively managed for the benefit of humans in different parts of the world. Some are set aside and managed solely on the basis of their scientific interest, and recreation is actively discouraged. In other areas, various recreational activities are compatible with the aims of the reserve. In the U.S.A., natural wetlands are accepted as having a wide variety of uses including nature study, photography, hunting, fishing, frogging, boating, camping and picnicking, as well as economic pursuits such as lumbering, haying, grazing, mining, petroleum extraction and fur-harvesting. While the primary interest in these areas may be wildlife or recreation, economic operations can yield substantial financial returns, and may even increase the value of the area for the wildlife or recreational activity concerned.

8.2.8 *Fisheries*

Though published figures for the annual world catch from fresh waters are always substantially less than those from the sea, the figures rarely include subsistence fishing or sport fishing, which in some countries at least (e.g. the U.S.A.) exceed the commercial catch from fresh waters. There is a remarkable similarity in the annual yield from marine continental shelf areas, which contribute about 80% of the total marine catch, and from fresh waters. The respective figures are 11 kg/ha and 12 kg/ha.

Fish farming and the production rates involved have been discussed above. Yields are greater than from wild populations in natural lakes and rivers, especially where fertilisers and feeding are utilised. All large lakes in the world have commercial fisheries, the average production from such waters being about 5 kg/ha/annum. The fifty largest fresh waters in the world occupy more than one-fifth of the total area of all inland waters and have a total area of more than 125 million hectares.

Some of the fish forming the basis of commercial fisheries are introduced, thriving at the expense of, or in a niche unoccupied by, native species. Successful examples of such introductions are whitefish in Lake Sevang in Russia, and Pacific salmon in the North American Great Lakes.

Most freshwater fisheries, however, rely on native species caught by methods which vary according to the species, water and tradition of local fisherpeople. In large lakes, fish are caught by gill nets and traps, though in some waters seine nets or trawls may be used. In temperate areas, salmon, trout, charr, whitefish, pike, perch, pikeperch and other species are important commercially, while in tropical regions other species are dominant. Most river fisheries rely on traps to capture both catadromous and anadromous species, though in some broad rivers and estuaries seine nets may be used. In temperate rivers, shad, salmon, trout, charr and eels are the main species caught, while in tropical running waters, *Vimba* and other lotic species are important.

8.3 Human influence

Only in recent years have humans become aware of the enormous damage being done to natural resources. Human influence on fresh waters is no exception. The conflict between the demand for large amounts of pure water on the one hand, and the disastrous pollution of many waters on the other hand, is only now forcing the issue with politicians. Multipurpose river basin projects seem the logical way to solve many problems. The following brief accounts indicate the major impacts of human usage on the ecology of freshwater systems.

Abstraction. The effect of this varies in extent, except with total abstraction where the results are obvious and disastrous. More often, only partial abstraction occurs, with variations in effect from year to year and place to place. In standing waters subject to rapid fluctuations in level caused by pumped-storage or flood-control projects, the shoreline experiences similar changes to those in abstracted rivers. There is a great reduction in macrophyte vegetation and in invertebrates which cannot withstand desiccation. Consequently, the shallow littoral areas of abstracted lakes and rivers, normally the richest zones, have poor production and specialised communities restricted to organisms which can withstand periodic desiccation, or are highly mobile and can keep pace with water level changes.

Impoundments. Provided there is an enlightened policy for river flow control, reservoirs can be beneficial within natural catchments. The result of building a dam across a tributary valley is obvious locally, and the flora and fauna of the reservoir undergo rapid change from a lotic to a lentic community, which may take some years to stabilise. The effect of reservoirs is not only local, however, and changes occur in the river system below. The outflow from a reservoir contains more plankton than the original stream, and stream animals feeding on them become commoner. Lakes also exert a stabilising influence on rivers (specifically so where water levels are controlled for flood prevention), and fluctuations in water level and temperature are reduced. In rivers with migratory fish, impoundments and weirs may affect ascent and descent; fish ladders and lifts are often built in to overcome this.

Hydro-electricity. Hydro-electric schemes have deleterious effects on local waters, because of abstraction, dams and turbines, water transfer and other activities. However, some fish are less affected than others by such schemes, and there is evidence that plankton-feeding fish may be favoured by the fluctuating water levels which affect the main feeding grounds of benthic feeders in the littoral zone. It is believed that, because fluctuating water levels devastate the littoral flora and fauna (Smith *et al.*, 1987), the benthic feeders, which occupy the littoral area, are adversely affected. Plankton is less affected and so plankton feeders still have their food source.

Drainage. Straightening and canalisation of river courses to prevent flooding may be done in a crude way, and often the river bed is so altered that it can never return to its original condition. Recolonisation by plants and animals from other areas may still be possible but, because of simplification of the channel environment, the decrease in microhabitats leads to impoverishment. Piping or ditching land to improve drainage increases the dangers resulting from higher water levels in wet weather and lower levels in dry weather, both a direct result of the faster run-off of water from the land (Stuart, 1959).

Agriculture. The clearing of forests also increases the run-off of surface water and the rate of soil erosion with subsequent silting and nutrient increase in the waters draining such areas. Most types of cultivation lead to loss of soil and nutrients; deficiency of the latter is commonly overcome by the regular addition of agricultural fertilisers. These too tend to be washed off and affect the nutrient status and ecology of waters into which they drain. Rapid eutrophication caused by increased nutrient input from fertilisers and sewage effluents is one of the major problems in the management of fresh waters today (Figure 8.7).

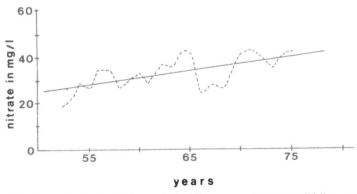

Figure 8.7 Increasing levels of nitrate (dashed line = annual means; solid line = trend) in the River Avon, England from 1950 to 1975, arising from the use of fertilisers in agriculture (after Royal Commission for Environmental Pollution, 1979).

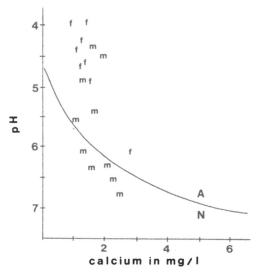

Figure 8.8 Comparisons of the pH and calcium in streams on moorland (m) and on similar ground recently afforested (f) (after Harriman and Morrison, 1982). The curve divides acidified (A) from non-acidified waters (N) (after Henriksen, 1979).

Afforestation. All stages of forestry—ground preparation, planting to the mature crop and felling—have impacts on fresh waters. Physical aspects affect: (1) stream hydrology, by (a) increased water loss through interception and evaporation from the forest, and (b) higher flood peaks and lower drought levels; (2) sedimentation of streams and lakes from erosion; (3) reduced summer water temperatures from tree shading. Chemical changes from afforestation include: (1) increased nutrients from leaching and fertilisers and (2) acidification from air pollutants (Figure 8.8) leading to high aluminium levels. These effects combine to affect freshwater plants and animals. Changes in hydrology and water temperatures make conditions more extreme for biota. Turbidity decreases plant growth and increased nutrients increase algae. Acidification affects plants and invertebrates and may eliminate fish. Amphibians and birds may be reduced in number or eliminated.

Fish farming. Fish farms pose a number of environmental problems, especially in clean-water areas where the industry is expanding rapidly (Figure 8.9). To combat disease, the use of disinfectants and antibiotics is commonplace. As well as direct chemical treatment, fish are treated by

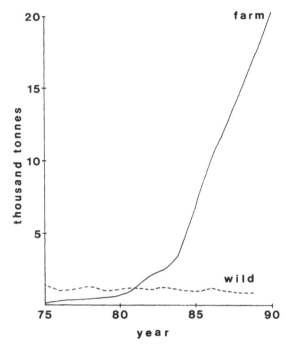

Figure 8.9   The increase in production of farmed Atlantic salmon in Scotland compared with the wild catch (after Maitland, 1989).

enteral administration of pharmaceutical products within feed (Schnick *et al.*, 1986). The amounts of such materials reaching natural waters are low, but do affect microflora (Austin, 1985). Solids from waste food and faeces pass into lakes or rivers, silting the bed and deoxygenating water. Nutrients leaching from fish feed, fish urine and faeces and from waste feed include nitrogen and phosphorus (Hansson, 1985) and typical amounts from salmonid farms average 92.75 and 18.75 kg/tonne of fish production per annum respectively, causing eutrophication and other problems. Fish farms can be a source of disease to wild fish; many fish farmers disclaim this, but they themselves are the first objectors to new fish farms near them. Many farmed fish find their way into local streams and lakes and interact with native fish by competing for space and food and predating eggs and young of native fish. Fish farms import various strains of fish from abroad and develop domestic races with characteristics unlikely to be advantageous in the wild. Fish are introduced to the wild in such numbers that they may upset the genetic integrity of native stock (Maitland, 1989).

Fisheries. The commercial harvesting of native freshwater fish usually has little harmful influence on the waters concerned. Indeed, because it is in their interests to avoid contamination, fisherpeople act as a strong force against pollution and other influences. Occasional harm may be done to fish populations by overfishing, or where poisoning is carried out to collect some species or to control undesirable types. Efficient sustainable cropping probably has relatively little effect on the system as a whole.

Angling. There has been increasing controversy in recent years concerning the impact of angling on aquatic wildlife. A central problem concerns litter, which, in addition to being unsightly, has a serious impact on birds and mammals because of hooks and monofilament line in which they become entangled. The presence of anglers often disturbs wildlife. Anglers can alter habitat, either unintentionally (e.g. by trampling down vegetation), or intentionally (e.g. weed cutting and bank clearance). Anglers may also impinge directly on aquatic communities by poisoning unwanted fish, shooting predatory birds or introducing new fish species. Stocking with desired fish species to enhance the population may have the opposite effect. The reasons for this are varied but include the introduction of disease, overstocking and the elimination of important genetic components evolved by the native stock.

Recreation. Much recreational use of fresh waters has little effect other than in certain special cases, and difficulties arising are often due to conflict between the different types of recreation involved rather than their influence on the aquatic system. However, most recreational uses of fresh waters (wildfowling, angling, sailing, bathing, boating and water-skiing) cause pollution and disturbance to certain animal species by actively killing them (wildfowl or fish) or frightening them away—an important problem with many nesting birds. Areas of shoreline and stands of macrophytic vegetation may be affected by trampling or the frequent passage of boats. Accumulations of lead from boats and oil from their motors are causing considerable pollution in many lakes.

Pollution. The influence of polluting substances on natural waters is variable according to local conditions and organisms within the water concerned. Pollutants can act in three main ways: by settling out on the substrate and smothering life there, by being acutely toxic and killing organisms directly, or by reducing the oxygen supply so much as to kill organisms indirectly. Since even clean cold water holds only about 12 mg/l oxygen, there is never a great deal available compared with air. Pollution may also be caused in other ways—by the addition of substances which may act as acids, alkalis or as nutrients. Radioactive substances, tainting of domestic water supplies, and alteration of water temperatures are other examples.

Effluents with high suspended solids are typical of mining industries, poorly treated domestic sewage, and various washing processes. Most of the solids settle out soon after discharge, at a rate dependent on their size, density and local current conditions. The effect of inorganic particles is mainly a physical one, and plants and invertebrates may be completely covered and destroyed. Fish often die through their gills becoming clogged. If the particles involved are organic, their decay may add the problem of deoxygenation to that of alteration of the substrate.

The impact of toxic substances on organisms in natural waters is complicated by the fact that different species have varying resistance thresholds to poisons (which may act variably at different temperatures) and that some poisons are cumulative in their effect and others are not. Most toxic substances originate from industrial processes, though some arise from mining and agriculture.

Organic materials in sewage effluents are a source of major pollution of fresh waters. Though these effluents often contain plant nutrients, these cannot be utilised for some time because of the high oxygen demand of the decomposing organic material. In extreme cases, especially in lakes and slow-flowing rivers, so much oxygen is used up that anaerobic conditions result and no organisms other than bacteria and some fungi can exist. In less severe cases, species with low oxygen requirements (e.g. tubificid worms and chironomid midge larvae) can exist, and indeed, in the absence of predators and competitors and with abundant supplies of organic material for food, may build up dense populations. It is common, especially in running-water situations, to find a sequence of changes from the area of greatest pollution near the effluent to cleaner water further away (Figure 8.10). With improving conditions as organic material is oxidised, there is a return to natural flora and fauna, though the quality and productivity may be influenced by the nutrient salts present.

Heated waters. With the high production of electricity from thermal power stations, the temperature regime of many natural waters has been significantly influenced by heated effluents. Relatively little is known about

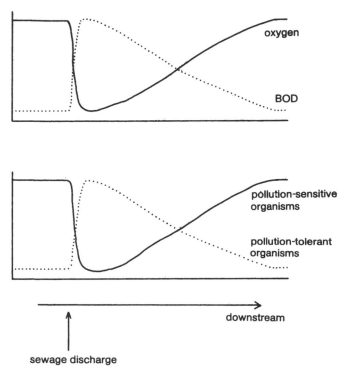

Figure 8.10   The impact of a bad sewage effluent on the oxygen and fauna of a river.

their influence on natural communities; it is likely that high temperatures will kill some stenothermic species and favour the development and reproduction of others. Also, the effect of organic effluents and some toxins is likely to be increased. In some temperate areas, species restricted naturally to tropical waters (e.g. the guppy *Poecilia reticulata*) have become established in the vicinity of heated effluents. The main effects of heated effluents as far as pollution is concerned are that warmer water holds less oxygen than cooler, and decomposition processes are speeded up.

Acidification.   Acid deposition, arising mainly from the burning of fossil fuels (e.g. in coal and oil power stations) has resulted in severe damage to fish in Canada, U.S.A., Scotland, Norway and Sweden. Salmonid fish are particularly vulnerable, but most fish have been affected. One of the characteristic features of acidification on fish is the failure of recruitment, manifested in an altered age structure and reduced population. This reduces intraspecific competition for food and increased growth or condition of survivors. As well as pH, the total ion content of the water is important to fish survival. There is concern for the future of many systems in the poorly buffered areas of the northern hemisphere if acidification continues.

8.4 Pollution prevention

Various definitions have been attached to the term pollution, but as it affects fresh waters it can be described as the discharge into a natural water of materials (usually waste products) which adversely affect the quality of plant and animal life there. Though some pollution does take place naturally, the most important contribution comes from human activities in three fields: agriculture, industry and domestic waste disposal. The importance of the interrelationship between fresh water and the disposal of human waste cannot be stressed too strongly. As human activities on earth increase, more and more fresh water is required; most of this is polluted before being returned to the water course, and in many cases a major use of the water is to flush away wastes. As more water is used, so are more natural waters polluted; but with the demand for water, more and more polluted systems are being utilised for water supply, thus involving elaborate and expensive purification plants. An understanding of pollutants, their treatment and their effect on freshwater systems is an important part of the interpretation of freshwater ecology today.

Though there are several ways of dealing with sewage, the most common system involves transport in a hydraulic system, after which it may be discharged direct to a water course or to the sea, or treated and then so discharged. In some sewer systems, liquid wastes and drainage water (from roads and rooftops) are combined, in others they are partially or completely separate. The latter system, though more expensive, is more efficient because smaller volumes of sewage need to be treated and there is no overflow during storms.

In this pollution-conscious era there is increasing pressure to treat sewage before discharge. The standard of effluent required depends on the capacity of the receiving water for self-purification and on the downstream use of water. Modern sewerage systems and sewage treatment plants are expensive, and the capital cost of their installation is the main factor against their development. Before the installation of new sewage

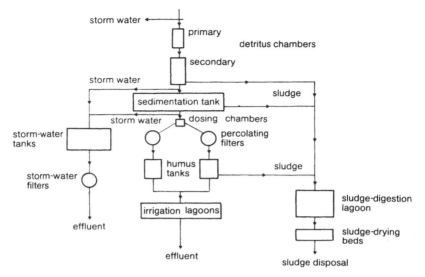

Figure 8.11   Lay-out of a standard sewage treatment plant.

plants (Figure 8.11) careful analyses should establish the quantity and quality of materials requiring treatment, and how these may vary in time. With this information it is then possible to decide on the type of treatment plant needed and the time required for its various activities, which may be considered in four stages: preliminary treatment, sedimentation, organic breakdown and final effluent treatment, followed by the disposal of the purified effluent.

Most sewage effluents are discharged directly into natural waters. Formerly it was rare for there to be restrictions on disposal into the sea, and so most of the sewage discharged in this way underwent little treatment. However, obvious pollution of the sea is now forcing authorities to consider treatment. Most inland discharges are to rivers, and in many countries their quality is controlled by law to ensure that at a minimum the water course remains aerobic and large sludge banks do not develop.

Though the wastes from many industries (e.g. dairies, abattoirs, canneries, tanneries and breweries) are troublesome because of their high organic content, they can be mixed with domestic sewage and treated conventionally. Other industries, such as mining and sand-quarrying, may produce effluents with large quantities of suspended inorganic solids which can be removed by special settling tanks. The most difficult wastes are those containing toxins (e.g. from paper-making, gas liquor and chemical trades); for obvious reasons these cannot be discharged into normal treatment systems depending on biological activity, and they must undergo special chemical treatment to remove the toxins.

8.5 Integration of water use

The combined effect of population growth, increased standards of living and greater leisure time is to place larger and larger demands on available water resources. In many areas these are already insufficient to meet present requirements. It is clear that increased cooperation regarding the integration of water use is essential at regional, national and international levels. Buchan (1963) has pointed out that water can be conserved only by interrupting the hydrological cycle to make more water available, or to make it available for longer periods. There are six main ways of achieving this:

(a) By diverting precipitation normally falling over the sea on to the land.
(b) By reducing evaporation rates over inland waters, especially from fresh waters themselves.
(c) By delaying run-off to the sea.
(d) By recirculating water used for cooling or other purposes.
(e) By purifying contaminated waste water so that it can be re-used.
(f) By desalinating brackish or sea water.

All these procedures are feasible, but may not be economic, especially if considered in isolation from other local water uses.

In several countries the concept of integrated water usage on the basis of large catchment areas has become accepted as the most rational approach to water conservation. In the United States, the Commission on Water Resources Policy recommended that new proposals for water resource development should be submitted only in the form of programmes which deal with entire river basins and which take into account all relevant features of water and land development. One of the first areas to develop its water resources under such a concept was the Tennessee Valley Authority, which evolved as a single coordinated project for that region. Within this programme there are now more than thirty reservoirs, over half of them including power, navigation and flood-control systems. Similar successful examples of multipurpose river basin development include the Snowy Mountain

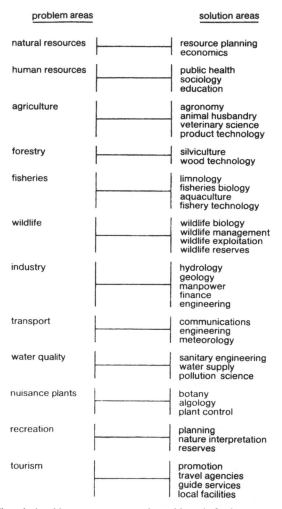

Figure 8.12 The relationships among man-made problems in fresh waters and the technical fields involved in solving them.

Authority in Australia, the Helmand Valley Authority in Afghanistan, and projects in South America and elsewhere (Figure 8.12).

As part of a project for water conservation in the Nile Basin in North Africa, the development of the enormous Aswan Dam (with a storage capacity of 130 000 million cubic metres) is giving the following benefits: large amounts of water for irrigation, cheap hydro-electricity, and conservation of water during heavy rains, thus improving flood control, drainage, agriculture and navigation further downstream. Unfortunately this project is also creating vast new difficulties, for instance the important fishery in the delta has virtually disappeared. The Nile lands were formerly very fertile from the nutrient-rich silt deposited periodically by the river; this silt is now retained and sedimented in the dam. The Damodar Valley project in India is an equally ambitious scheme with several large reservoirs which, as well as providing flood control, increased navigation facilities and hydro-electricity, means a great increase in the local fish crops and in the area under irrigation.

All these schemes and many smaller ones developing in some parts of the world are based largely on economic reasoning. Nevertheless, if planned and implemented intelligently, they may lead to improved amenities from both recreational and aesthetic viewpoints—especially in the long term. The rational use of freshwater resources leading to improved standards of living, clean lakes and rivers and increased recreational facilities can only be commended.

CHAPTER NINE
A GLOBAL VIEW

For complex geological, climatic and other reasons there are enormous differences between the fresh waters of different parts of the world. This is true also of marine communities, but especially so of freshwater ones where isolation has tended to emphasise the role of historical events. There is a very evident increase in the general abundance and diversity of freshwater plants and animals from the poles to the tropics, but the most striking and important differences are among the major continents of the world (Table 9.1) and even between major river basins in some continents. Unfortunately, progress in the conservation of fresh waters in different parts of the world bears little relation to the relative importance of the indigenous aquatic communities there.

Several zoogeographers have emphasised that many groups of freshwater organisms are uniquely significant in zoogeography. Those that occur only in fresh water are as closely bound to the major land masses as are any land animals and they are normally inescapably confined to their own drainage systems, only passing from one basin to the next by chance mechanisms or the intervention of humans. Thus each continent, perhaps each basin, can have a unique freshwater community, and freshwater ecology viewed on a world scale must take these important geographical considerations into account.

Lake (and, to a lesser extent, river) faunas are similar in many ways to those of islands—they are isolated, small and relatively young (Magnuson, 1976). Thus theories developed for the population dynamics of island populations (MacArthur and Wilson, 1963) should be relevant to lakes. The number of species in a lake may be the result of equilibrium between rates of immigration of new species and extinction of old ones. The number of species in small isolated lakes would be lower than in large more complex ones or those more exposed to new species. Instability in species structure would be higher in the latter. The influence of immigration

Table 9.1 Details of the area (km^2, 10^3), annual rainfall (mm), evaporation and transpiration (as percent rainfall) and run-off (as percent rainfall) from the major continents (after Australian Water Resources Council, 1976).

Continent	Area	Rainfall	Evaporation	Run-off
Africa	30 210	660	76	24
Asia	44 030	610	64	36
Australia	7 690	420	87	13
Europe	9 710	580	60	40
North America	24 260	660	60	40
South America	17 790	1350	64	36

on species structure would be greater in smaller and simpler habitats than in larger and more complex ones. Human activities in moving aquatic organisms around the world can increase instability and add greater uncertainty to the future of aquatic communities.

Although the bulk of freshwater systems eventually connect to one of the great oceans of the world, many do not, and these are termed areas of inland drainage. Virtually all the inland drainage areas are situated in two great belts around the globe, one in the northern hemisphere approximately between latitudes 30 and 60° North, and the other in the southern hemisphere between latitudes 20 and 30° South. These regions correspond closely to the great desert areas of the world. Many of the standing waters are salt lakes where there is an annual rainfall of less than 25 cm and the relationship is obviously one of cause and effect (Murray and Pullar, 1910). In the northern belt are the lakes of the Gobi Desert, the Aral, Caspian and Dead Seas, the lakes of the Arabian and Sahara Deserts and the Great Salt Lake and alkali deserts of North America. In the southern belt are the arid regions of the interior of Australia, the Kalahari Desert of Africa and the Atakama Desert of South America. These inland drainage areas are estimated to occupy some 30 million km^2 or about 18% of the total land surface of the globe.

The primary cause of these arid areas with inland drainage and many salt lakes depends more on meteorology than topography or geology. They arise mainly because they are situated on those parts of the earth's surface where the prevailing winds blow from colder to warmer latitudes and from off land, not directly off the ocean.

Fish are particularly important in relation to the zoogeography of fresh waters, and faunas of primary freshwater fishes are confined to the continents (and certain continental islands)—excepting Australia which has virtually no primary freshwater species (Darlington, 1980). This suggests that many of these fish have been confined to fresh water for a very long time.

9.1 Fresh waters around the world

9.1.1 *Africa*

Because of the low rainfall and the dryness of the Saharan area, most of the important fresh waters of Africa occur south of the Sahara. There are many enormous and very old lakes on this continent, notably Lake Chad (10–25 000 km^2) and the great lakes of east Africa—Lake Victoria (68 800 km^2), Lake Tanganyika (34 000 km^2), Lake Malawi (30 800 km^2) and Lake Turkana (6410 km^2). Some very large and important rivers occur here also—the Nile (6695 km), Zaire (4667 km), Niger (4100 km), Zambezi (2655 km), Orange (1859 km) and Gambia (1094 km). Most of these rivers and many of the shallower lakes experience annual or biannual fluctuations in flow or area due to the seasonal pattern of rainfall (Table 9.2).

The dynamics of most of the rivers and to a lesser extent the lakes are also dictated by the seasons. Conductivity is lowest at high water and vice versa. In some inland areas evaporation is so great that alkaline systems have developed. Many of the waters are rich and their phytoplankton is dominated by blue–green algae. An enormous variety of invertebrate species occur thoughout African freshwater systems, but many of these species are very little known.

Africa possesses a diverse freshwater fish fauna in which cyprinids, characins and catfishes are abundant. The communities are richest and

Table 9.2 Areas of peak flood (km^2) around some major tropical rivers.

River	Area	Authority
Amazon	50 000	Sioli (1975)
Magdalena	20 000	Pardo (1976)
Niger	20 000	Raimondo (1975)
Nile	92 000	Rzoska (1974)
Okavango	17 000	Welcomme (1979)
Orinoco	70 000	Welcomme (1979)
Paraguay	80 000	Bonetto (1975)
Parana	20 000	Bonetto (1969)
Senegal	5 000	Welcomme (1979)
Zambezi	10 752	FAO (1969)

most diverse in West Africa, but the great lakes of east-central Africa have highly endemic fish faunas dominated by cichlids—many of these are under severe threat at the moment (Barel *et al.*, 1985). The great richness and diversity of ancient stocks is unequalled anywhere else in the world. These occur from the west tropical area to the Nile and are richest of all in the Congo. Faunas in the east are moderately poor, there is local radiation of cichlids in the great lakes and progressive poverty in diversity in a southwards direction. The northern desert area is depauperate and the north-western corner has a limited but different fauna closely related to that of Europe.

9.1.2 *Antarctica*

The huge continent of Antarctica, though of major importance in many ways, is of relatively little importance in a freshwater context. Most of the continent is permanently frozen—as are the fresh waters. Some have many metres of permanent ice over a few metres of water and others may be permanently frozen to the bottom. Even those areas to the north which thaw out regularly have either no land mass or reveal only small lakes (which may have no outflow to the sea) and temporary streams.

These extreme systems have extremely impoverished floras (mainly algae, lichens and mosses) and faunas (notably protozoans and rotifers), with no freshwater fish. Most of the organisms are capable of being frozen into the ice for years at a time.

9.1.3 *Asia*

Asia has an enormously extensive and varied land mass with a wide variety of fresh waters, especially large rivers which flow to the Arctic Ocean in the north, the Pacific Ocean in the east and the Indian Ocean in the south. One of the largest and most important standing waters in the world—Lake Baikal—occurs here; this is a deep long trough 30 500 km^2 in surface area. It is the deepest lake in the world (1649 m) and its bed is 1165 m below sea level. Its water is clear and transparent and temperature is constant below 300 m. Its flora and fauna have many endemic forms (some of them close to marine groups) including molluscs, crustaceans, fish and seals.

Other large lakes in Asia include Lake Balkhash (17 400 km^2) in Russia and Tung-ting Lake, which is the largest lake in China (5003 km^2) with a very variable water level. Poyang Lake (4250 km^2), which also fluctuates greatly in size, is a regulator for the Yangtze-Kiang River (5794 km)—Asia's

longest running water. The many other large rivers of Asia include the
Ob (5150 km), Huang He (4840 km), Amur (4667 km), Irtysh (4440 km),
Lena (4281 km), Mekong (4180 km), Yenisei (4506 km), Brahmaputra
(2960 km), Salween (2820 km) and Ganges (2510 km). The range of habitats
and plant and animal communities within this vast resource is enormous.

In Asia, the main characteristics of the fish fauna are a richness (in spite
of the absence of archaic groups) in tropical Asia, with poverty and speciali-
sation in the central highlands, a transition through eastern Asia from the
tropics north, and a wide east–west distribution of northern groups with
evolution of endemic groups in Lake Baikal.

9.1.4 *Australia*

Australia is an extremely dry continent, with little rain and high rates of
evaporation. Thus very little of the water falling on the land actually ends
up running into the sea. Williams (1981) has pointed out that the total
annual run-off in Australia is about 346 km³—only slightly in excess of
the volume discharged annually by one European river, the Danube!
Thus in many places, especially in the west, the running waters are
temporary. Most of the permanent rivers are in the eastern half of the
continent and these are characterised by wide seasonal and annual variations
in flow.

Most inland waters are slow-flowing, except in the mountains and near
the coast. Because of the irregularity of the rainfall in many parts of
Australia, many of the rivers may at times be dry for months, but at other
times rushing torrents carrying down great amounts of silt which are
deposited in the sluggish lower reaches. Many of the waters are rather
turbid and often dominated chemically by calcium and magnesium. Waters
in the centre of the continent are typically saline and impermanent.

There is a large inland drainage basin near the centre of the continent,
and various temporary waters flow into this. Lake Eyre (8294 km²) is a
large inland water within this basin which was originally fresh and much
larger but is now smaller and salty. Its actual size in any one year varies
enormously according to rainfall.

The most distinctive feature of the fauna of Australian fresh waters is
its high endemicity. This is clearly reflected in the aquatic invertebrates,
for example the various endemic species of mussels, crustaceans, stoneflies
and fish. Among the crustaceans are the unique members of the Anaspidacea,
while among the fish the large family Galaxiidae has many representatives
found only in Australia. Even the few aquatic mammals are quite unique—

notably the platypus, *Ornithorhynchus*, the only known aquatic monotreme (Figure 2.9(8)).

Apart from lungfish and osteoglossids, Australia has considerable numbers of peripheral freshwater fish, particularly lampreys, galaxiids, catfishes, perch-like fish, gobies and atherinids. Some of these native species are widely distributed over Australia. Others (e.g. lampreys and galaxiids) are confined to the cooler southern parts of the continent. The fish fauna of New Guinea (except possibly osteoglossids) is entirely peripheral, and as a whole is closely related to that of Australia.

9.1.5 *Europe*

Europe has a relatively restricted range of geographic area and climate, and its fresh waters and their communities are correspondingly less varied than those of most other continents. There are numerous small and medium-sized lakes, but only a few very large ones. These include Lake Ladoga, which is the largest lake in Europe (18 182 km², maximum depth 223 m). It is only 5 m above the Gulf of Finland, with which it was once connected. Like many European waters it freezes every winter, and ice masses can reach over 20 m in height. The nearby Lake Onega (9600 km²) is the second largest European lake, and Lake Vanern (5580 km²) in Sweden is next in size. Europe has many important rivers, but none of them approach the enormous running waters of Asia and America. Among the largest are the Volga (3688 km), Danube (2850 km), Dnepr (2285 km), Don (1870 km), Pechora (1799 km), Dnestr (1410 km) and Rhine (1320 km).

Between Europe and Asia stretches a large inland drainage area from *ca* 35 to 125° east and from 25 to 60° north and includes the lakes of the Aralo-Caspian depression, of the Gobi Desert, Lake Hamun and Lake Urmi in Iran and Lake Van in Turkey. The Caspian and Aral Seas are salt lakes which owe their saltness to having originally been part of the ocean from which they were separated by movements of the earth's crust. Molluscs and other animals in the Caspian Sea are very similar to those of the Black Sea, to which it was at one time probably connected. The Caspian Sea is the largest inland body of water in the world—some 371 000 km², with a maximum depth of 975 m. The deepest waters are anaerobic with little life, but a high proportion of the fauna is endemic, including species of seals and sturgeons. The Aral Sea is another large brackish water which was once connected to the Caspian but is now about 75 m above it.

In Europe the fish fauna is distinctly poor—increasingly so to the west and in Iberia and in Italy, but with a few endemic and/or relict groups in the south-east. The poverty in the south is compensated for in part by the presence of various peripheral fish of Mediterranean origin. Northern Europe shares the greater part of the widely distributed northern Asiatic fauna.

9.1.6 North America

This huge continent has an enormous range of fresh waters draining in all directions and covering a wide range of geography and habitat. Many of the waters are large, notable among which are the Great Lakes (Superior, Huron, Michigan, Erie and Ontario), five enormous inter-connected waters which eventually drain east to the Atlantic Ocean via the St Lawrence River. They are among the largest lakes in the world (Lake Superior is the largest) and their dimensions are included in Table 3.1. As well as these waters, there are other large lakes, including Great Bear ($31\,330\,km^2$), Great Slave ($28\,570\,km^2$) and Winnipeg ($24\,390\,km^2$).

There are many major rivers too, including the Mississippi ($6260\,km$), Mackenzie ($4240\,km$), Missouri ($3969\,km$), St Lawrence ($3058\,km$), Rio Grande ($3030\,km$) and Yukon ($3020\,km$). The inland drainage areas of North America cover over $700\,000\,km^2$ and an important water within these is Great Salt Lake, which is $1280\,m$ above sea level and $5000\,km^2$ (but variable) in area.

Among the fish communities of North America there is richness and diversity in the east (especially in the Mississippi basin) which decreases along the Atlantic coast and especially northward into Canada, and, to a lesser extent, southward into Mexico. West of the continental divide the fauna is mainly cyprinids, catostomids and cyprinodonts with some peripheral fishes near the coast. The transition northwards is into an arctic fauna of a few primary divisions and many peripheral fishes related to those of northern Asia and Europe.

9.1.7 South America

The drainage pattern of South America is dominated by one enormous drainage system—that of the River Amazon. This is the largest river in the world with a length of $6274\,km$, an average discharge of $172\,400$ cumecs and draining a catchment of $6\,133\,000\,km^2$. However, there are several other substantial rivers in the continent, notably the Parana (4500 km), Madeira (3240 km), Sao Francisco (2780 km), Orinoco (2060 km) and Negro (2000 km).

There are no very large lakes in South America comparable to those on other continents, although a number of medium-sized standing waters occur. The inland drainage areas are estimated at some 1 300 000 km². Much of this drainage accumulates in two lakes—Titicaca (fresh and deep) and Poopo (shallow with salt water). The two are remnants of a once vast inland sea but are now connected by a running water (Desaguardero). Lake Titicaca, which has a surface area of 8294 km² and a maximum depth of 270 m, is one of the highest lakes in America.

The freshwater fish fauna of South America is notable for its small number of ancestral stocks but the richness and endemicism of various groups of Ostariophysi. Thus of primary freshwater fish only five species are not in this group. In the Ostariophysi there are really only characins, gymnotids and catfishes—but these number altogether about 2000 known species. The main freshwater fish fauna of this continent centres on the basin of the River Amazon—by far the richest river system in the world in terms of fish species diversity. This diversity declines in adjacent river systems and especially towards the south, where there is a transition to a completely different fauna of antarctic peripheral fishes.

9.2 International problems

9.2.1 Translocations

The ways in which freshwater plants and animals disperse across land masses have always been an area of debate. Hypotheses concerning natural phenomena which enable stenohaline organisms to cross from one watershed to another include inter-catchment connections, whirlwinds and waterspouts, migrating aquatic birds and other means. Some scientists doubt the importance of natural means and believe that humans are the main agents in moving aquatic organisms overland. Whatever the contribution of the different factors in the past it certainly seems true that humans are now the major force in moving species around the world. The ecological consequences of some of these translocations have been catastrophic in a number of cases, but there are also examples where the results are believed to have been beneficial.

Plants and invertebrates, fish and other vertebrates have been moved around the world on thousands of occasions, sometimes intentionally, sometimes unintentionally. Most aquatic species have been moved during the last century and many even during the last few decades. The case of Canadian pondweed has already been discussed (page 159), but other case histories are instructive in relation to such species and their impact on the waters into which they have been introduced.

The Chinese mitten crab *Eriocheir sinensis* is, as its name implies, a native of China, but appeared in Europe in the River Weser in 1912, having probably travelled from the East in the ballast tanks of ships (Gledhill *et al.*, 1976). By 1927 it was widespread in Germany, occurring in rivers many hundreds of kilometres from the sea and had become a pest because of its habit of burrowing into river banks. It is now found over much of Europe.

At Loch Ness in Scotland the presence of large numbers of *Phagocata woodworthi*—a species of North American triclad new to Britain—was of considerable interest (Reynoldson *et al.*, 1981). It seems likely that this species was introduced into the loch by American teams looking for the 'Loch Ness Monster' by eggs or adults having been transported over the Atlantic Ocean on or in equipment, some of which was known to have been tested in North American waters. This represents a casual or unintentional method of introduction, but one which could have serious consequences in some circumstances.

Maitland and Price (1969) found that a population of the North American largemouth bass, *Micropterus salmoides*, naturalised in a pond in Dorset was host to the monogenetic trematode parasite *Urocleidus principalis*. This parasite, which is specific to the genus *Micropterus*, had never before been recorded in Europe but was known as a common parasite of largemouth bass in North America. It was assumed that the parasite came to Great Britain with its host and became established there with it.

One of the most notable disease problems in Europe in recent years has been the outbreak in Norway of the parasitic fluke *Gyrodactylus salaris* which has been spread by introductions from farmed salmonids to wild populations, and several native stocks of salmon have been virtually wiped out (Dolmen, 1987). Over the last decade it has proved necessary to use poison to eradicate entire fish stocks and communities in some rivers as the only means of eliminating the parasite. The original infection appears to have arisen from the import of parasitised stock from fish farms in Sweden, and the parasite is now known from twenty-eight rivers and eleven hatcheries in Norway.

The history of the accidental introduction of the sea lamprey into the

Figure 9.1   Whitefish (*Coregonus*) from one of the Great Lakes of North America, showing the massive wounds inflicted by sea lampreys (*Petromyzon*) (Photo: U.S. Fish and Wildlife Service.)

Great Lakes of North America has been well documented. The opening of the Welland Canal in 1829 gave this species access to the upper Great Lakes for the first time, but it was not found in Lake Erie until 1921, Lake Huron until 1937, Lake Michigan until 1936 and Lake Superior until 1946. In the next few decades the population of lamprey developed explosively with catastrophic consequences for several of the fish species on which it preyed (Figure 9.1), notably American lake charr *Salvelinus namaycush* (Figure 9.2), lake whitefish *Coregonus clupeaformis* and lake herring *Coregonus artedii*. The commercial fisheries for these species virtually collapsed and Smith (1968) estimated that in Lake Michigan alone during the mid-1950s the sea lamprey destroyed some five million pounds of fish a year.

The Great Lakes Fishery Commission was set up to study this and other problems and it has developed a series of lamprey-control strategies including electrical and physical barriers to prevent adult lampreys ascending their spawning streams and chemical lampricides in the nursery streams to eliminate larvae. The latter technique has proved the most

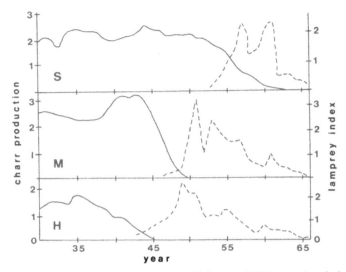

Figure 9.2 Production (catch) of lake charr (solid lines in 10 000 tonnes) and abundance of sea lamprey (dashed lines) in the upper Great Lakes (S = Superior; M = Michigan; H = Huron) of North America (after Smith, 1968).

successful and is the main thrust of the control programme costing, during the 1980s, some $8 000 000 per annum.

In Lake Wingra, Wisconsin, U.S.A., there is considerable information on species numbers and human influence (Baumann *et al.*, 1974). During the 10 000 years prior to 1900, fourteen fish became established. In the next fifty years, species numbers doubled—mainly from introductions. The number is now declining due to intense species interactions and some habitat modification (e.g. the introduction of the plant *Myriophyllum* from Europe). The fish community has changed from one dominated by pike and perch to one dominated by stunted centrarchids and yellow bass *Morone mississipiensis*. There are many other cases of harmful fish introductions in North America (e.g. Lemly, 1985).

Not all introductions are regarded as having been harmful, at least in fishery terms. The introduction of rainbow smelt *Osmerus mordax* into Schoodic Lake, Maine, U.S.A., as a new forage base for landlocked salmon resulted in a marked increase in mean salmon size within three years of the initial smelt introduction in 1965 (Havey, 1973). The 3 + salmon in 1965–67 averaged 364 mm and 468 g, but in 1968–70 they averaged 442 mm and 814 g. The condition factor in 1966 was 0.76; in 1969 it was

1.00. A drastic decline in the population of smelt in the winter of 1971 seriously curtailed salmon growth.

Altukhov (1981) has presented genetic evidence to show that even when wild stocks of the same species are transferred from one river to another damage can occur. Wild stock of the chum salmon *Oncorhynchus keta* in the River Kalininka in Sakhalin (Russia) was considered to be 'superior' to that in the nearby River Naiba, and from 1964 to 1971 over 350 million eggs were transferred from the Kalininka to the Naiba. Instead of improving the situation, the results were disastrous (Figure 9.3). Only about 10–20% of the expected number of Kalininka fish returned initially and the returns of local fish were reduced. The population fell from 650 000 spawners in 1968 to about 35 000 in 1980, and by 1985 the population was virtually extinct.

Ben-Tuvia (1981) reviewed the situation in Lake Kinneret in Israel, where thirteen exotic species and two native to other parts of Israel have been introduced, some intentionally, others accidentally. He concluded that the introduction of exotic species has 'proved to be economically advantageous'. However, this conclusion was rebutted by Gophen *et al.*

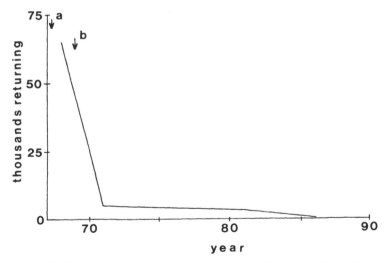

Figure 9.3 Decline of the spawning chum salmon population in the River Naiba after massive transplants from the River Kalininka (after Altukhov, 1981; and Maitland, 1989). a indicates when 350 million eggs were transferred from the Kalininka; b indicates when the Naiba stock started showing genetic features of the Kalininka stock.

(1983) who produced contrary evidence and concluded: 'We ask the simple question. Which should be more effective in developing a stable and productive system: Man and his introduction of exotic species or 20 000 years of Lake Kinneret evolution? We suggest the latter.'

The Chinese grass carp has been introduced to a number of waters in Great Britain as a method of biological control of water weeds and to provide sport fishing. The initial introductions were carried out on an experimental basis (Stott, 1977) and the research showed that this fish was efficient at controlling weed, was popular with anglers and was never likely to breed naturally in Great Britain. The Chinese grass carp has also been widely distributed in many parts of the world as a food fish and for weed control. There has been considerable controversy over its introduction to the United States, where some states have banned it whilst others have encouraged its introduction. The uncertainty arose in part because of the damage caused earlier to many habitats by the introduction of common carp *Cyprinus carpio*—originally hailed as a useful introduction (Black, 1946).

In Great Britain, the ruffe *Gymnocephalus cernus* is indigenous to the south-east of England from where it has spread via canal systems to the English Midlands and eastern parts of Wales. The previous most northerly record appears to have been from the River Tees, and the species never seems to have been recorded from Scotland or North America. In 1982, ruffe appeared in Loch Lomond 100 km north of its former area of distribution (Maitland *et al.*, 1983) and it is now one of the commonest fish in the loch, and may have ousted perch *Perca fluviatilis* to some extent (Figure 9.4). It is believed that the ruffe was introduced to Loch Lomond by anglers from England, who frequently fish for pike with small fish as live bait. The impact of this new species on the existing community is uncertain, but unlikely to be beneficial, especially to the vulnerable powan *Coregonus lavaretus*, whose eggs it eats in large numbers at spawning time.

Perhaps even more surprising than this move within Britain is the enormous move that the ruffe has recently made from Europe to North America. In 1987, several ruffe were found in the St Louis River where it enters Lake Superior (McInnes, 1988). It is now well established in this part of the lake. It is believed that the original fish were brought accidentally from Europe in the ballast water of seagoing freighters travelling from one of the large rivers of Europe to the Great Lakes. The Great Lakes Fishery Commission is actively trying to bring in legislation which will force such ships to change their ballast water in mid-ocean or

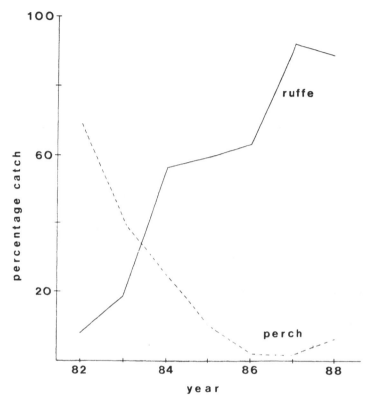

Figure 9.4 Numbers of ruffe and perch in catches of fish from Loch Lomond, expressed as a percentage of the total fish catch from the screens of a water supply pumping station on the loch shore.

treat it in some way to stop any further exotic organisms entering North America in this way.

These various cases indicate that tighter international controls should exist for at least certain classes of aquatic plants and animals as well as perhaps for procedures by which organisms could be translocated accidentally. These include not only making sure that fish and other organisms imported are disease- and parasite-free, but also that other routes are blocked. For example, no ship should be allowed to move into fresh water in any country with fresh water from abroad as ballast. It should be illegal to use in fresh water equipment or anything else that has been used in fresh water abroad without first sterilising it in some way.

Some years ago, the International Council for the Exploration of the Sea adopted a code of practice to reduce the risks arising from the introduction of non-indigenous marine species (ICES, 1973). A similar code was adopted for inland waters by the European Inland Fisheries Advisory Commission, and recently the American Fisheries Society and the International Council for the Conservation of Nature and Natural Resources have issued positive statements on the question of introductions and transfers. There have also been a number of positive proposals from individual scientists (e.g. Ryder and Kerr, 1984).

9.2.2 *Acidification*

Although known to be a problem since the 1920s, it is only over the last fifteen years or so that ecologists in the northern hemisphere have become increasingly concerned about the impact of acid deposition on freshwater ecosystems and other parts of the environment (Almer *et al.*, 1974; Beamish *et al.*, 1975; Harvey, 1975; Haines, 1981). In Scandinavia (especially southern Sweden and Norway), and more recently in North America (in both Canada and the U.S.A.), numerous scientific studies related to the problem have been initiated and there was a massive growth in the literature during the 1980s. Many other countries are now involved in work on acid deposition, but in Great Britain the input in the field of freshwater ecology has been relatively small until recently (United Kingdom Acid Waters Review Group, 1986).

Considerable general agreement appears to be developing from the research data (involving field survey, monitoring and experimental work) produced in Scandinavia and North America (Drablos and Tollan, 1980; Haines, 1981; Johnson, 1982). Rain in many parts of the world—including the British Isles (United Kingdom Review Group on Acid Rain, 1986; Mathews *et al.*, 1984)—is acid and has probably become more so during this century. This rain and associated dry deposition appears to have acidified some fresh waters—especially those in areas of base-poor geology whose buffering capacity is low.

Organisms at each major trophic level are affected by this acidification. The diversity of phytoplankton decreases with acidification, but the production of some algae and mosses increases (Battarbee, 1984). The diversity and production of most macrophyte communities decreases with decreasing pH, and the same appears to be true of zooplankton and zoobenthos, though the situation is more complex with invertebrates (Engblom and Lingdell, 1983). If the acidification is sufficiently great to

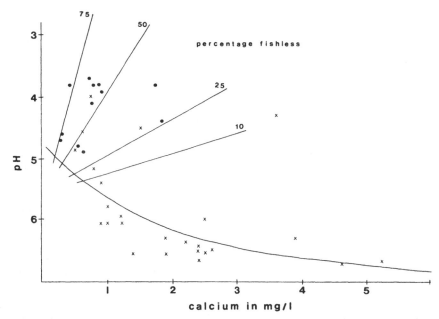

Figure 9.5 pH and calcium plots of a range of Scottish lochs superimposed on the acidification curve of Henriksen (1979) and hypothesised proportions likely to be fishless, calculated from the data of Wright and Snekvic (1978) (after Maitland *et al.*, 1987).

exclude fish, then their absence as the normal top predators can lead to an unusual abundance of some prey species. Amphibians (Tome and Pough, 1982) and birds (Eriksson, 1984) can also be affected.

One of the earliest indicators of acid pollution and one of its most important effects was the disappearance of many fish (Figure 9.5)— especially salmonids (Atlantic salmon *Salmo salar*, brown trout *Salmo trutta* and arctic charr *Salvelinus alpinus*)—from rivers and lakes in which they were previously abundant (Wright & Snekvic, 1978; Muniz & Leivestad, 1980; Harvey & Lee, 1982; Schofield, 1982). For example, Atlantic salmon have disappeared from many rivers in southern Scandinavia, and the number of lakes in these areas without populations of brown trout and arctic charr has increased dramatically, especially over the last fifteen years (Overrein *et al.*, 1980; Johnson, 1982). Massive kills of salmon and trout have been observed during snowmelt and after heavy rain (Henriksen *et al.*, 1984).

One of the most characteristic features of acidification on fish populations is the failure of recruitment of new age classes into the population (Harvey, 1982; Rosseland *et al.*, 1980). This is manifest in an altered age structure and reduction in population size (Figure 9.6). This reduces intraspecific competition for food, where this resource is limiting, and increases growth or condition of survivors. Beamish and Harvey (1972) have shown that the disappearance of fish during the acidification process often follows a pattern. As a lake is subjected to a long-term decrease in the pH of its water the normally large numbers of small fish disappear at some critically acidic level. The fish stock then starts to 'improve' and there are lower numbers of fine large fish. However, with no recruitment the population contains fewer and fewer ageing fish until eventually there are no fish at all (Figure 9.6).

As well as pH, the total ion content of the water is important to fish survival (Brown, 1982). In natural waters, higher concentrations of ions increase survival time under otherwise similar conditions—including pH. Several workers have tried to define and classify various levels of acidification and the resultant fish communities. Kelso and Minns (1981) have produced such a scheme (Table 9.3), which seems to indicate that there should be some concern for the future of many systems in the poorly buffered areas of the northern hemisphere if acidification continues there. Arctic charr and trout are among the least tolerant of freshwater species (Table 9.4) (Almer *et al.*, 1974).

Both population size and size distribution of fish have been known to change dramatically in acid lakes following neutralisation. During the late 1960s and early 1970s, the pH of the Swedish lake Stora Skalsjon was 4.5–5.5 (Harvey, 1982). It was limed in 1976 and 1977, gradually raising the pH to 6.0. The catch of perch per net set increased in 1978 and 1979, concurrent with the reintroduction and expansion of arctic charr. In the short term, it is probable that the main way of ameliorating the effects of acidification in aquatic systems is by the use of lime, and already many thousands of lakes in Scandinavia have been treated in this way.

9.2.3 *Global warming*

Global warming, due to the so-called 'greenhouse effect', is a recently perceived phenomenon which may eventually have an enormous influence on the world, including its freshwater ecosystems. There is now general agreement among scientists—climatologists in particular—that global warming has already started. Accumulating in the atmosphere, the increasing amounts, created by human activities, of 'greenhouse gases'

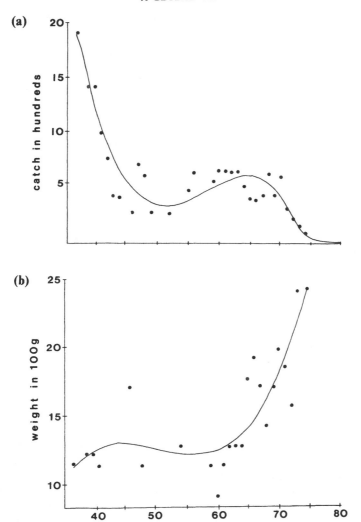

Figure 9.6 Reduction in angler catch (a) and increasing weight (b) of trout taken in Loch Grannoch, an acidifying loch in south-west Scotland (after Hay, 1984).

Table 9.3 Classification of lake systems and their fish in relation to acidification (after Kelso and Minns, 1981).

Category	Conductivity	pH	Risk to fish
A		< 4.7	Fishless
	30		
B		4.7–5.3	Remnant fish population
	50		
C		5.3–6.0	High risk
	70		
D		6.0–6.5	Moderate risk

Table 9.4 Tolerance of various fish species to low pH (after Almer *et al.*, 1974).

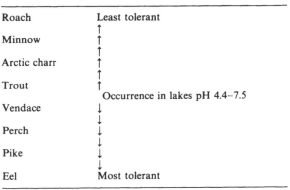

Roach	Least tolerant
Minnow	↑
Arctic charr	↑
Trout	↑
	Occurrence in lakes pH 4.4–7.5
Vendace	↓
Perch	↓
Pike	↓
Eel	Most tolerant

(carbon dioxide (Figure 9.7), chlorofluorocarbons, methane and nitrous oxide) trap the radiant heat of the sun and cause a general warming of the atmosphere (Figure 9.8). As temperatures rise, glaciers melt in the great mountain ranges of the world and their waters run to the oceans increasing levels there. The oceans themselves warm and expand, melting oceanic ice towards the poles and thus further increasing sea levels.

Since the last ice age, some 20 000 years ago, the sea has been rising—but very slowly. Maximum annual rises seem to have been about 1–2 cm a year, and over the last century levels have risen by only about 1.2 mm a year. It is generally expected that global warming will accelerate this rate of increase, but there is some debate as to the actual extent of the increase. Early forecasts predicted a rise of about 65 cm by the year 2030 (this would

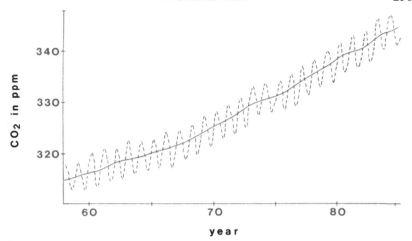

Figure 9.7 Increasing quantities of carbon dioxide (dashed line = annual variations; solid line = annual mean) recorded in Hawaii over a period of three decades (after Mauna Loa Observatory, 1985).

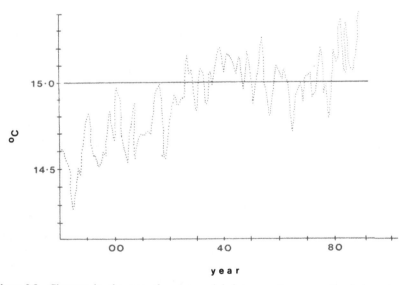

Figure 9.8 Changes in the annual average global temperature over the last century (1880–1988). The straight line indicates the average temperature for the years 1950–80.

give sea levels about 3.5 m higher than present by the twenty-second century), but the latest estimates suggest a rise of 15–30 cm by 2030, and this seems a realistic view on present evidence.

The prognosis for some parts of the world is extremely worrying, especially low-lying areas close to the sea. Among the most threatened areas are the low-lying coral islands of the Indian and Pacific Oceans; most of these are only a few metres above sea level and many are densely populated. Therefore there is no area for the population to retreat to, and their way of living (which is largely dependent on the resources of the shallow seas around them) is threatened. Several other low-lying countries (e.g. Bangladesh and Guyana) already have substantial problems from regular flooding caused by heavy inland rains or coastal storms or a combination of both. Their difficulties will be exacerbated as global warming takes effect.

It seems therefore that there will be world-wide changes in climate and sea level which are likely to affect most habitats and their communities in some way. Freshwater ecosystems will be no exception.

On a small scale, one of the effects of global warming is already evident and has been studied to some extent (Milner, 1989). In parts of Alaska, U.S.A., there has been a retreat of some glaciers and ice sheets over the last few decades, leaving behind barren areas of land and clear, cold, virgin streams. These are colonised at various rates by several invertebrates (e.g. stoneflies) and fish (e.g. the dolly varden). Most of the initial invertebrate colonisers are insects occurring in waters to the south, and these obviously have little difficulty colonising such waters during their aerial phase. The fish which move into such streams and establish themselves are all anadromous species, and these too have an obvious route of invasion from the sea.

In many countries all over the world, one of the most important areas to be flooded as the sea rises will be estuaries and their deltas. Not only are some of the systems involved very extensive (e.g. the Ganges), but these areas are frequently the focus of much human activity and habitation, as well as diverse habitats for wildlife. There is no doubt that permanent flooding of estuaries will cause human problems on an enormous scale. The effect on ecosystems is less certain, for many of the organisms there are used to sudden changes of temperature, salinity and substrate. Therefore, it seems likely that though most estuaries will deepen and increase in size, the organisms there will be able to adapt to changing circumstances and redistribute themselves accordingly.

The flooding of the low-lying flood plains inland is another matter as far as local ecosystems are concerned, especially where there is an incursion of salt water. Here, the invasion of the estuary into the former flood plain waters will eliminate most of the existing flora and fauna, which will be replaced by estuarine species. It seems certain that much of the lentic zone of many large rivers will be replaced by extended estuary, and in some places it may be eliminated altogether if the estuary reaches as far inland as the lotic zone. Thus many existing lowland rivers will be reduced or eliminated and their floras and faunas with them. For instance, lowland fish communities which include, in Europe, such fish as bream, tench and roach may be significantly affected and rivers will tend more towards some of those already common in northern Scotland (e.g. the River Dee) where there is virtually no lentic section and the highland lotic zone flows more or less directly into the sea.

It could be postulated from this that highland lotic species (e.g. most salmonids) may fare better than lowland lentic ones (e.g. many cyprinids) as global warming proceeds and lowland lakes and rivers are flooded by rising sea levels. However, it must be remembered that at the same time ambient temperatures will be rising and these may well affect the ecology of cold-loving species, which many northern aquatic fish and other organisms are. Thus it would seem likely that the distribution of such species will gradually change and there will be a tendency for a retreat to higher latitudes and altitudes, where conditions will be more suitable.

To give an example of this: at present the mayfly *Ameletus inopinatus*, which is found in southern Britain, occurs only in mountainous areas (e.g. the English Lake District) and not below about 300 m. Further north (see Figure 4.10), it is found at lower and lower altitudes until in northern Scotland it occurs right down near sea level. It can be assumed that species such as this will be eliminated from parts of their present southern distribution and forced into higher altitudes in northern areas. Similarly, and perhaps more important, salmonid fish will be affected in the same way and are likely to show a retreat in their distribution towards the north, with extinctions in a number of southern habitats. To an extent these will be compensated in some areas by their occupation of new waters, as discussed above.

These are just a few of the possible scenarios if global warming proceeds according to present expectations. Considerable concern now exists among the major nations of the world, and possible solutions to the problem are being proposed. One of the main contributors to the 'greenhouse effect' is the enormous amount of carbon (estimated at some five billion tonnes annually) being released into the atmosphere (Figure 9.7). Extensive forest plantations have been proposed to absorb this carbon by photosynthesis. However, other solutions have been put forward, and one of these concerns aquatic systems (Toha and Jaques, 1989). The proposal, which involves the mass cultivation of algae, is said to ten times more efficient than planting forests and could be implemented immediately. Given adequate nutrients and average climate, algal concentrations of $0.3 \, g/l$ can be achieved, with a doubling time of three days. Thus a 1 ha pond with a depth of 1 m would absorb 1800 kg of carbon daily. Approximately twenty-two million hectares of such ponds would be needed to absorb the excess carbon dioxide in the atmosphere today and it is suggested that these could be distributed along the coastline of some sixty countries. Some of the details of such proposals are arguable, but they do help to illustrate the enormity of the problem and the scale at which it will need to be tackled if the present theories about global warming are correct.

9.3   The future

One of the most important revolutions which is still taking place world-wide is the pressure for change in land use. This has been brought about for a variety of reasons, partly the financial advantages in planting some crops (e.g. conifers) and partly (in some developed countries) the recent over-production of foodstuffs. There are likely to be enormous changes in land use over the next few decades and these will undoubtedly have effects on freshwater ecosystems. Some of these effects may be good, others bad.

The main trend in many developed countries is likely to be a change away from conventional and intensive agriculture towards other forms of land use, particularly forestry and leisure activities. Overall, this is likely to be beneficial to the freshwater environment, especially in the lowland areas, where a reduction in fertilisers, herbicides, pesticides and less intense drainage with more ground under both deciduous and coniferous trees should all prove beneficial. In the uplands, however, increasing coniferous afforestation is already creating problems and in some base-poor heavily afforested areas the waters have become acidic and completely fishless.

Aquatic communities face a number of problems, some of them common to other forms of wildlife, others more particular to fresh waters. Rivers and to a lesser extent lakes are repositories of enormous amounts of human waste, ranging from toxic industrial chemicals through agricultural slurries and herbicides to domestic sewage. As discussed above, even aerial pollutants such as sulphur dioxide from power station chimneys are eventually washed into water courses as 'acid rain'.

Many rivers have become completely fishless as a result, especially those in the industrial and heavily populated lowland areas. Other factors have affected fresh waters in various ways. Barriers on rivers, such as weirs or hydro-dams have blocked the passage of migratory fish to their spawning grounds and so eliminated them. Enrichment from farm fertilisers, overfishing and the introduction of exotic species have all contributed to the decline of some communities, especially those with rare and more sensitive native species. Aquatic populations are mostly limited by land boundaries to their immediate water body and thus the whole population is vulnerable to a single incident of toxic spillage or acidification.

Over the last few decades there have been very significant advances in combating pollution in some countries. The trend towards increasingly polluted waters and declining fish populations has been reversed and many rivers are now much cleaner than they were fifty or even a hundred years ago. Some rivers which were so badly polluted that they became fishless are now clean again and supporting good stocks of fish. The Rivers Clyde in Scotland and Thames in England are good examples.

There is no doubt that public support and feeling behind the worldwide conservation movement has strengthened substantially over the last decade. In many countries people are supportive of wildlife conservation in some form and have an image of what that means. Natural habitats (meadows and mountains, forests and seashore, rivers and lakes), and wildlife (trees and flowers, butterflies and bees, birds and mammals) are areas of popular interest and concern. However, many aquatic species are themselves difficult to observe in the wild and so do not yet have a similar popular following. Yet there is substantial interest among some groups of people in living fish and other aquatic animals. There are many millions of anglers throughout the world, but sadly their main concern is with the species they wish to catch, and indeed some do harm in a number of ways, such as moving fish around and introducing them as predators or competitors to the waters containing rare species. Another major interest group is aquarists, but often these are uninterested in native fish and are concerned mainly with exotic species.

There are now many nature reserves of various types managed by different organisations throughout the world. They have been established for numerous different reasons; some for a unique type of habitat or plant community, many for their ornithological interest and others for the rare flowers or butterflies which occur there. Few reserves have been set up especially for their fish or other aquatic life but hopefully this is something which will be remedied in the future.

Eventually, through the work of aquatic ecologists and enthusiastic conservationists we will understand much more about the status and requirements of aquatic species and communities, and more of them will be given protection in national parks or nature reserves. The need is clearly a long-term one. Hopefully, by the end of the century, much ground (really water!) which has been lost will have been recovered and more and more fresh waters will be safe for future generations to use and enjoy in various ways.

REFERENCES

Aitken, P.L. (1963) Hydro-electric power generation. *Proc. Symp. Inst. Civil Eng. Lond.* **1963**, 34–42.

Alabaster, J.S. and Lloyd, R. (1980) *Water Quality Criteria for Freshwater Fish.* Butterworth, London.

Allen, K.R. (1951) The Horokiwi Stream: a study of a trout population. *N. Z. Mar. Dept. Fish. Bull.* **10**, 1–231.

Almer, B., Dickson, W., Ekstrom, C. Hornstrom, E. and Miller, U. (1974) Effects of acidification on Swedish lakes. *Ambio* **3**, 330–336.

Altukhov, Y.P. (1981) The stock concept from the viewpoint of population genetics. *Can. J. Fish Aquat. Sci.* **38**, 1523–1538.

Ambuhl, H. (1959) Die Bedeutung der Stromung als ekologischer Faktor. *Schweiz. Z. Hydrol.* **21**, 133–264.

American Public Health Association (1960) *Standard Methods for the Examination of Water and Wastewater,* American Public Health Association, New York.

Andersson, F. and Olsson, B. (1985) *Lake Gardsjon: an Acid Forest Lake and its Catchment,* Swedish Research Councils, Stockholm.

Atkins, W.R.G. (1945) Daylight and its penetration into the sea. *Trans. Illum. Eng. Soc.* **10**, 1–12.

Austin, B. (1985) Antibiotic pollution from fish farms: effects on aquatic microflora. *Microbiol. Sci.* **2**, 113–117.

Australian Water Resources Council (1976) *Review of Australian Water Resources,* Australian Government Public Service, Canberra.

Awachie, J.B.E. (1981) Running water ecology in Africa. In Lock, M.A. and Williams, D.O., *Perspectives in Running Water Ecology,* Plenum, New York, 339–366.

Axelrod, H.R. (1980) *Cardinal Tetras,* Neptune, New York.

Bagenal, T.B. (1969) The relationship between food supply and fecundity in brown trout *Salmo trutta* L. *J. Fish. Biol.* **1**, 167–182.

Bailey-Watts, A.E. and Duncan, P. (1981) Chemical characterisation. A one year comparative study, *Monogr. Biol.,* **44**, 67–89.

Balvay, G. (1967) L'oxygene dissous dans le lac d'Annecy, a la fin de la stagnation estivale. *Bull. Soc. Hist. Nat. Macon.* **5**, 7–9.

Bamforth, S.S. (1958) Ecological studies on the planktonic Protozoa of a small artificial pond. *Limnol. Oceanogr.* **3**, 398–412.

Barel, C.D.N., Dorit, R., Greenwood, P.H., Hughes, N., Jackson, P.B.N., Kawanabe, H., Lowe-McConnell, R.H., Nagoshi, M., Ribbink, A.J., Trewavas, E., Witte, F. and Yamaoka, K. (1985) Destruction of fisheries in Africa's lakes. *Nature, Lond.* **315**, 19–20.

Barnes, H. (1959) *Oceanography and Marine Biology,* Macmillan, New York.

Barnes, R.D. (1968) *Invertebrate Zoology,* Saunders, Philadelphia.

Battarbee, R.W. (1984) Diatom analysis and the acidification of lakes. *Phil. Trans. Roy. Soc. Lond.* **305**, 193–219.

Bauman, P.C., Kitchell, J.F. and Magnuson, J.J. (1974) Lake Wingra 1837–1973. A case history of human impact. *Wisc. Acad. Sci. Arts. Lett.* **62**, 57–94.

Beadle, L.C. (1943) Osmotic regulation and the faunas of inland waters. *Biol. Rev.* **18**, 172–183.

Beadle, L.C. (1974) *The Inland Waters of Tropical Africa*, Longman, London.

Beamish, R.J. and Harvey, H.H. (1972) Acidification of the La Cloche Mountain Lakes, Ontario, and resulting fish mortalities. *J. Fish. Res. Bd. Can.* **29**, 1131–1143.

Beamish, R.J., Lockhart, W.L., Van Loon, J.C. and Harvey, H.H. (1975) Long term acidification of a lake and resulting effects on fishes. *Ambio* **4**, 98–102.

Bellairs, A.D. (1957) *Reptiles*, Hutchinson, London.

Ben-Tuvia, A. (1981) Man-induced changes in the freshwater fish fauna of Israel. *Fish. Mgt.* **12**, 139–145.

Berg, K. (1948) Biological studies on the River Susaa. *Folia. Limnol. Scand.* **4**, 1–318.

Berrie, A.D. (1970) Snail problems in African schistosomiasis. *Adv. Parasitol.* **8**, 43–96.

Beverton, R.J.H. and Holt, S.J. (1957) On the dynamics of exploited fish populations. *Fish. Invest., Lond.* **10**, 1–533.

Bindloss, M.E. (1974) Primary productivity of phytoplankton in Loch Leven, Kinross. *Proc. Roy. Soc. Edin.* **74**, 157–181.

Birge, E.A. and Juday, C. (1922) The inland lakes of Wisconsin. The plankton. I. Its quality and chemical composition. *Wisc. Geol. Nat. Hist. Surv.* **64**, 1–222.

Birge, E.A. and Juday, C. (1932) Solar radiation and inland lakes. Fourth report: observations of 1931. *Trans. Wisc. Acad. Sci. Arts Lett.* **27**, 523–562.

Birge, E.A. and Juday, C. (1934) Particulate and dissolved organic matter in inland lakes. *Ecol. Monogr.* **4**, 440–474.

Black, J.D. (1946) Nature's own weed killer – the German carp. *Wisc. Cons. Bull.* **11**, 3–7.

Blum, J.L. (1956) The ecology of river algae. *Bot. Rev.* **22**, 291–341.

Bonetto, A.A. (1969) Limnological investigations on biotic communities in the middle Parana River valley. *Verh. Int. Ver. Limnol.* **17**, 1035–1050.

Bonetto, A.A. (1975) Hydraulic regime of the Parana River and its influence on ecosystems. *Ecol. Stud.* **10**, 175–197.

Borgstrom, G. (1971) *Fish as Food*, Academic Press, New York.

Boughey, A.S. (1968) *Ecology of Populations*, Macmillan, New York.

Brinkhurst, R.O. (1975) *The Benthos of Lakes*, Macmillan, London.

Brinkhurst, R.O. and Jamieson, B.G.M. (1971) *Aquatic Oligochaeta of the World*, Oliver & Boyd, Edinburgh.

Brook, A.J. and Woodward, W.B. (1956) Some observations on the effects of water inflow on the plankton of small lakes. *J. Anim. Ecol.* **25**, 22–35.

Brown, C.J.D. and Flaton, C.M. (1943) A portable field chemistry kit. *Limnol. Soc. Amer., Spec. Publ.* **11**, 1–4.

Brown, D.J.A. (1982) The effect of pH and calcium on fish and fisheries. *Water Air Soil Pollut.* **18**, 343–351.

Brown, D.J.A. and Sadler, K. (1989) Fish survival in acid waters. *Soc. Exp. Biol. Semin. Ser.* **34**, 31–44.

Bruce, J.P. and Clark, R.H. (1966) *Introduction to Hydrometeorology*, Pergamon, Oxford.

Buchan, S. (1963) Conservation by integrated use of surface and ground water. *Proc. Symp. Inst. Civil Eng. Lond.* **1963**, 181–185.

Butcher, R.W. (1933) Studies on the ecology of rivers. I. On the distribution of macrophytic vegetation in the rivers of Britain. *J. Anim. Ecol.* **212**, 58–91.

Butcher, R.W., Pentelow, F.T.K. and Woodley, J.W.A. (1927) The diurnal variation of the gaseous constituents of river waters. *Biochem. J.* **2**, 945–957.

Cadwalladr, P.L. and Backhouse, G.N. (1983) *A Guide to the Freshwater Fish of Victoria*, Government Printer, Victoria.

Campbell, R.M. (1961) The pattern of existing water use. In Elgood, L.A., *Natural Resources in Scotland*, Constable, Edinburgh.

Carpenter, K.E. (1927) Faunistic ecology of some Cardiganshire streams. *J. Ecol.* **15**, 33–54.

Carpenter, K.E. (1928) *Life in Inland Waters*. Sidgwick & Jackson.

Carr, J.F., Moffat, J.W. and Gannon, J.E. (1973) Thermal characteristics of Lake Michigan, 1954–55. *Tech. Pap.. Bur. Sport Fish Wildl.* **69**, 1–143.

Chester, P.F. (1984) Chemical balance in lakes in Sorlandet. *Proc. Roy. Soc. Lond. B*, **305**, 564–565.

Chrystal, G. (1910) Seiches and other oscillations of lake surfaces observed by the Scottish Lake Survey. *Bath. Surv. Freshw. Lochs Scot.* **1**, 29–90.

Clark, D. and England, G. (1963) Thermal power generation. *Proc. Symp. Inst. Civil Eng. Lond.* **1963**, 43–61.

Clarke, F.W. (1924) The composition of river and lake waters of the United States. *U.S. Geol. Surv. Pap.* **1924**, 1–135.

Clarke, G.L. (1939) The utilisation of solar energy by aquatic organisms. *Amer. Ass. Adv. Sci. Publ.* **10**, 27–38.

Clarke, G.L. and Bumpus, G.F. (1950) The plankton sampler—an instrument for quantitative plankton investigations. *Limnol. Soc. Amer. Spec. Publ.* **5**, 1–8.

Coker, R.E. (1954) *Lakes, Streams and Ponds*, University Press, North Carolina.

Copeland, J.J. (1936) Yellowstone thermal Myxophyceae. *Ann. New York Acad. Sci.* **36**, 1–232.

Corliss, J.O. (1961) *The Ciliated Protozoa: Characterisation, Classification and a Guide to the Literature*, Pergamon, New York.

Cott, H.B. (1940) *Adaptive Coloration in Animals*, University Press, Oxford.

Cunningham, W.A. (1910) On the nature and origin of freshwater organisms. *Bath. Surv. Freshw. Lochs Scot.* **1**, 354–373.

Darlington, P.J. (1980) *Zoogeography: the Geographic Distribution of Animals*, Krieger, Huntington.

Davies, H.S. (1938) Instructions for conducting stream and lake surveys. *U.S. Dept Comm. Bur. Fish.* **26**, 1–55.

Denny, P. (1972) Lakes of southwestern Uganda. I. Physical and chemical studies of Lake Bunyoni. *Freshw. Biol.* **2**, 143–158.

Dolmen, D. (1987) *Gyrodactylus salaris* (Monogenea) in Norway; infestations and management. In Stenmark, A. and Malmberg, P., *Parasites and Diseases in Natural Waters and Aquaculture in Nordic Countries*, University of Stockholm, Stockholm, 63–69.

Douglas, B. (1958) The ecology of the attached stream diatoms and other algae in a small stony stream. *J. Ecol.* **46**, 295–322.

Drablos, D. and Tollan, A. (1980) *Ecological Impact of Acid Precipitation*, SNSF Project, Oslo.

Duffey, E. (1962) The Norfolk Broads. A regional study of wildlife conservation in a wetland area with high tourist attraction. *Project MAR, I.U.C.N. Publ.* **3**, 290–301.

Duncan, U.K. (1959) *A Guide to the Study of Lichens*, Buncle, Arbroath.

Edmondson, W.T. (1959) *Freshwater Biology*, Wiley, New York.

Edmondson, W.T. and Winberg, G.G. (1971) *A Manual on Methods for the Assessment of Secondary Productivity in Fresh Waters*, Blackwell Scientific Publications, Oxford.

Edwards, R.W. (1958) The effect of larvae of *Chironomus riparius* Meigen on the redox potentials of settled activated sludge. *Ann. Appl. Biol.* **46**, 457–464.

Edwards, R.W. and Owens, M. (1962) The effects of plants on river conditions. IV. The oxygen balance of a chalk stream. *J. Ecol.* **50**, 207–220.

Egglishaw, H.J. and Shackley, P.E. (1971) Suspended organic matter in fast-flowing streams in Scotland. I. Downstream variations in microscopic particles. *Freshw. Biol.* **1**, 273–285.

Elgood, L.A. (1961) *Natural Resources in Scotland*, Constable, Edinburgh.

Elliott, J.M. (1971) Some methods for the statistical analysis of samples of benthic invertebrates. *Sci. Publ. Freshw. Biol. Ass.* **25**, 1–144.

Elton, C.E. (1927) *Animal Ecology*, Sidgwick and Jackson.

Engblom, E. and Lingdell, P.E. (1983) Usefulness of the bottom fauna as a pH indicator. *SNV Publication* **1741**, 1–181.

Eriksson, M.O.G. (1984) Acidification of lakes: effects on waterbirds in Sweden. *Ambio* **13**, 260–262.

Fager, E.W. (1957) Determination and analysis of recurrent groups, *Ecology*, **38**, 586–595.

FAO (1969) *Report to the Government of Zambia on Fishery Development in the Central Barotse Floodplain*. Food and Agricultural Organisation of the United Nations, Rome.

Finlay, B.J. (1985) Nitrate respiration by Protozoa (*Loxodes* spp.) in the hypolimnetic nitrate maximum of a productive freshwater pond. *Freshw. Biol.* **15**, 333–346.

Finlay, B.J., Fenchel, T. and Gardiner, S. (1986) Oxygen perception and oxygen toxicity in the freshwater ciliated protozoan *Loxodes*. *J. Protozool.* **33**, 157–165.

Fittkau, E.J. (1964) Remarks on limnology of central-Amazon rain-forest streams. *Verh. Int. Ver. Limnol.* **15**, 1092–1096.

Fogg, G.E. (1968) *Photosynthesis*, English University Press, London.

Fogg, G.E., Stewart, W.D.P. and Walsby, A.E. (1973) *The Blue–green Algae*, Academic Press, London.

Forel, F.A. (1904) *Le Leman*, Monographie Limnologique. Lausanne.

Foskett, A.C. (1967) *A Guide to Personal Indexes*, Bingley, London.

Fritsch, F.E. (1929) Encrusting algal communities of certain streams. *New Phytol.* **28**, 165–196.

Fritsch, F.E. (1945) *Structure and Reproduction of the Algae*, University Press, Cambridge.

Fryer, G. (1959) Some aspects of evolution in Lake Nyasa. *Evolution* **13**, 440–451.

Geitler, L. and Ruttner, F. (1935) Die Cyanophyceen der Deutschen Limnologischen Sunda-expedition. *Arch. Biol. Suppl.* **14**, 308–715.

Gerking, S.D. (1967) *The Biological Basis of Freshwater Fish Production*, Blackwell Scientific Publications, Oxford.

Gilson, H.C. (1964) Lake Titicaca. *Verh. Int. Ver. Limnol.* **15**, 112–127.

Gledhill, T., Sutcliffe, D.W. and Williams, W.D. (1976) Key to British freshwater Crustacea: Malacostraca. *Sci. Publ. Freshw. Biol. Ass.* **32**, 1–72.

Goldman, C.R. (1966) *Primary Productivity in Aquatic Environments*, University of California Press, Berkeley.

Golterman, H.L. (1971) *Methods of Chemical Analysis for Fresh Waters*, Blackwell Scientific Publications, Oxford.

Gophen, M., Drenner, R.W. and Vinyard, G.G. (1983) Fish introduction into Lake Kinneret— call for concern. *Fish. Mgt.* **14**, 43–45.

Gorham, E. (1961) Factors influencing supply of major ions to inland waters, with special reference to the atmosphere. *Geol. Soc. Amer. Bull.* **72**, 795–840.

Griffin, A.E. (1963) Multi-purpose river basin development. *Proc. Symp. Inst. Civil Eng. Lond.* **1963**, 151–154.

Hack, J.T. (1957) Studies of longitudinal stream profiles in Virginia and Maryland, *U.S. Geol. Surv. Prof. Pap.* **294B**, 1–97.

Haines, T.A. (1981) Acidic precipitation and its consequences for aquatic ecosystems: a review. *Trans. Amer. Fish. Soc.* **110**, 669–707.

Hansson, S. (1985) Effects of eutrophication on fish communities with special reference to the Baltic Sea—a literature review. *Rep. Inst. Freshw. Res. Drottning.* **62**, 36–56.

Hardy, A.C. (1959) The continuous plankton recorder. *Discovery Rep.* **11**, 457–510.

Harriman, R. and Morrison, B.R.S. (1982) Ecology of streams draining forested and non-forested catchments in an area of central Scotland subject to acid precipitation. *Hydrobiologia* **88**, 251–263.

Harrison, A.D. and Elsworth, J.F. (1958) Hydrobiological studies on the Great Berg River, Western Cape Province. *Trans. Roy. Soc. S. Afr.* **35**, 125–329.

Harvey, H.H. (1975) Fish populations in a large group of acid stressed lakes. *Verh. Int. Ver. Limnol.* **19**, 2406–2417.

Harvey, H.H. (1982) Population responses of fish in acidified waters. *Proc. Int. Symp. Acid Precip. Cornell.* **1981**, 227–242.

Harvey, H.H. and Lee, C. (1982) Historical fisheries changes related to surface water pH changes in Canada. In Johnson, R.E., *Acid Rain: Fisheries*, Cornell University, New York.

Hasler, A.D. (1947) Eutrophication of lakes by domestic drainage. *Ecology* **28**, 383–395.

Havey, K.A. (1973) Effects of a smelt introduction on growth of landlocked salmon at Schoodic Lake, Maine, *Trans. Amer. Fish. Soc.* **102**, 392–397.

Hay, D. (1984) Acid rain—the prospect for Scotland. *Proc. Ann. Study Course, Inst. Fish. Mgt.* **15**, 110–118.

Henriksen, A. (1979) A simple approach for identifying and measuring acidification of fresh water. *Nature, Lond.* **278**, 542–545.

Henriksen, A., Skogheim, O.K. and Rosseland, B.O. (1984) Episodic changes in pH and aluminium speciation kill fish in a Norwegian river. *Vatten* **40**, 255–260.

Hickling, C.F. (1960) *Tropical Inland Fisheries*, Longmans Green, London.

Hickling, C.F. (1962) *Fish Culture*, Faber & Faber, London.

Hillebrand, D. (1950) Verkrautung und Abfluss, *Besond, Mitt. Dt. gewasserk. Jb.*, **2**, 1–30.

Holden, A.V. (1966) A chemical study of rain and stream waters in the Scottish highlands. *Freshw. Salm. Fish. Res. Scot.* **37**, 1–17.

Holmes, R.W. and Widrig, T.M. (1966) The enumeration and collection of marine phytoplankton. *J. Cons. Int. Explor. Mer.* **22**, 21–32.

Holt, S.J. (1967) In Gerking S.D., *The Biological Basis of Freshwater Fish Production*, Blackwell Scientific Publications, Oxford.

Hopthrow, H.E. (1963) Utilisation of water in industry. *Proc. Symp. Inst. Civil Eng. Lond.* **1963**, 30–33.

Huet, M. (1972) *Textbook of Fish Culture*, Fishing News, London.

Hunter, W.R., Maitland, P.S. and Yeoh, P.H.K. (1963) *Potamopyrgus jenkinsi* in the Loch Lomond area, and an authentic case of passive dispersal. *Proc. Malac. Soc. Lond.* **36**, 27–32.

Hutchinson, G.E. (1938) On the relation between the oxygen deficit and the productivity and topology of lakes. *Int. Rev. Hydrobiol.* **36**, 336–355.

Hutchinson, G.E. (1941) Ecological aspects of succession in natural populations. *Amer. Nat.* **75**, 406–418.

Hutchinson, G.E. (1957) *A Treatise on Limnology*, Wiley, New York.

Hutchinson, G.E. and Loffler, H. (1956) The thermal classification of lakes. *Proc. Nat. Acad. Sci. Wash.* **42**, 84–86.

Hyman, L.H. (1959) *The Invertebrates. V. Smaller Coelomate Groups*, McGraw-Hill, NY.

Hynes, H.B.N. (1958) The effect of drought on a small mountain stream in Wales. *Verh. Int. Ver. Limnol.* **13**, 826–833.

Hynes, H.B.N. (1959) The use of invertebrates as indicators of river pollution. *Proc. Linn. Soc. Lond.* **170**, 165–169.

Hynes, H.B.N. (1960) *The Biology of Polluted Waters*, University Press, Liverpool.

Hynes, H.B.N. (1970) *The Ecology of Running Waters*, University Press, Liverpool.

Hynes, H.B.N. (1973) The effects of sediment on the biota in running water, *Proc. Can. Hydrol. Symp.*, **9**, 652–663.

ICES, (1973) Code of practice to reduce risks of adverse effects from the introduction of non-indigenous marine species. *Proc. Reun. Cons. Int. Explor. Mer* **1973**, 50–51.

Illies, J. (1952) Die Molle, Faunistisch-okologische Untersuchungen an einem Forellenbach in Lipper Bergland. *Arch. Hydrobiol.* **46**, 424–612.

Isaac, P.C.G. (1953) *Public Health Engineering*, Spon, London.

Jackson, D.F. (1968) *Algae, Man and the Environment*, University Press, Syracuse.

Jayaram, K.C. (1981) *The Freshwater Fishes of India, Pakistan, Bangladesh, Burma and Sri Lanka*, Zoological Society of India, Calcutta.

Johannsen, O.A. (1932) Ceratopogonidae from the Malayan Subregion of the Dutch North East Indies, *Arch. Hydrobiol. Suppl.* **9**.

Johnson, R.E. (1982) *Acid Rain: Fisheries*, Cornell University, New York.

Jones, J.R.E. (1949) An ecological study of the river Rheidol, North Cordiganshire, Wales. *J. Anim.* **18**, 67–88.

Jones, J.R.E. (1951) An ecological study of the River Towy. *J. Anim. Ecol.* **20**, 438–450.

Jordan, M.J. (1985) Muror Lake–biological considerations. In: Likens, G.E. (ed.) *An Ecosystem Approach to Aquatic Ecology*. Springer-Verlag, New York, pp. 156–310.

Juday, C. (1940) The annual energy budget of an inland lake. *Ecology* **21**, 438–450.

Juday, C. (1942) The summer standing crops of plants and animals in four Wisconsin lakes. *Trans. Wisc. Acad. Sci. Arts Lett*, **34**, 103–135.

Kelso, J.R.M. and Minns, C.K. (1981) Current status of lake acidification and its effect on the fishery resources of Canada. *Proc. Int. Symp. Acid Precip. Cornell* **1981**, 69–90.

Kimball, J.W. (1965) *Biology*, Addison-Wesley, Reading.

Kimerle, R.A. (1968) Production biology of *Glyptotendipes barbipes* (Staeger) (Diptera: Chironomidae) in a waste stabilisation lagoon. PhD Thesis, Oregon State University.

Kimerle, R.A. and Anderson, N.H. (1971) Production and bioenergetic role of the midge *Glyptotendipes barbipes* (Staeger) in a waste stabilisation lagoon. *Limnol. Oceanogr.* **16**, 646–659.

King, H.W., Wisler, C.O. and Woodburn, J.G. (1948) *Hydraulics*, New York.

Klein, L. (1957) *Aspects of River Pollution*, Butterworths, London.

Kofoid, C.A. (1908) The plankton of the Illinois River, 1894–1899. III. Constituent organisms and their seasonal distribution. *Bull. Ill. State Lab. Nat. Hist.* **8**, 1–354.

Kolkwitz, R. and Marsson, M. (1909) Okologie der Flerischen Saprobien. *Int. Rev. Hydrobiol.* **2**, 126–152.

Krogh, A. (1939) *Osmotic Regulation in Aquatic Animals*, University Press, Cambridge.

Kuznetsov, S.I. (1958) A study of the size of bacterial populations of organic matter formation due to photo- and chemosynthesis in water bodies of different types. *Verh. Int. Ver. Limnol.* **13**, 156–169.

Lagler, K.F. (1949) *Studies in Freshwater Fishery Biology*, Ann Arbor.

Lagler, K.F. (1969) *Man-made Lakes*, FAO, Rome.

Lagler, K.F., Bardach, J.E. and Miller, R.R. (1962) *Ichthyology*, Wiley, New York.

Langford, R.R. (1953) Methods of plankton collection and a description of a new sampler. *J. Fish. Res. Bd. Can.* **10**, 238–252.

Langmuir, I. (1938) Surface motion of water induced by wind. *Science* **87**, 119–123.

Laybourn-Parry, J., Olver, J., Rogerson, A. and Duverge, P.L. (1989a) The temporal and spatial patterns of protozooplankton abundance in a eutrophic temperate lake. *Hydrobiologia*, in press.

Laybourn-Parry, J., Olver, J. and Rees, S. (1989b) The hypolimnetic protozoan plankton of a eutrophic lake. *Hydrobiologia*, in press.

Le Cren, E.D. and Holdgate, M.W. (1962) *The Exploitation of Natural Animal Populations*, Blackwell Scientific Publications, Oxford.

Lehninger, A.L. (1965) *Bioenergetics*, Benjamin, New York.

Lemley, A.D. (1985) Suppression of native fish populations by green sunfish in first-order streams, Piedmont, N. Carolina. *Trans. Amer. Fish. Soc.* **114**, 705–712.

Lin, S.Y. (1949) *Pond Culture of Warm Water Fishes*, UNESCO, Warm Springs.

Lindeman, R.L. (1942) The trophic dynamic aspect of ecology. *Ecology* **23**, 399–418.

Lloyd, D. (1942) Evaporation loss from land areas. *Trans. Inst. Wat. Eng.* **47**, 59–74.

Lock, M.A. and Williams, D.D. (1981) *Perspectives in Running Water Ecology*, Plenum, New York.

Lund, J.W.G. (1950) Studies on *Asterionella formosa* Hass. II. Nutrient depletion and the spring maximum. *J. Ecol.* **38**, 1–35.

Lund, J.W.G. (1965) The ecology of freshwater plankton. *Biol. Rev.* **40**, 231–293.

Lund, J.W.G. and Talling, J.F. (1957) Botanical limnological methods with special reference to the algae. *Bot. Rev.* **23**, 489–583.

Luther, H. and Rzoska, J. (1971) *Project AQUA: A Source Book of Waters Proposed for Conservation*, Blackwell Scientific Publications, Oxford.

Macan, T.T. (1958a) Methods of sampling the bottom fauna in stony streams. *Mitt. Int. Ver. Limnol.* **8**, 1–21.

Macan, T.T. (1958b) The temperature of a small stony stream. *Hydrobiologia* **12**, 89–106.

Macan, T.T. (1961) A review of running water studies. *Verh. Int. Ver. Limnol.* **14**, 587–602.

Macan, T.T. (1963) *Freshwater Ecology*, Longman, London.

Macan, T.T. (1970) *Biological Studies of the English Lakes*, Longman, London.

Macan, T.T. and Worthington, E.B. (1951) *Life in Lakes and Rivers*, Collins, London.

MacArthur, R.H. and Wilson, E.O. (1963) An equilibrium theory of island biogeography. *Evolution* **17**, 373–387.

McDowall, R.M. (1980) Interactions of the native and alien faunas of New Zealand and the problem of fish introduction. *Trans. Amer. Fish. Soc.* **97**, 1–11.

MacFadyen, A. (1957) *Animal Ecology: Aims and Methods*, Pitman, London.

McInnes, C. (1988) River ruffe called threat to fishery. *Globe & Mail* (May 3), 3.

Mackereth, F.J.H. (1963) Some methods of water analysis for limnologists. *Sci. Publ. Freshw. Biol. Ass.* **21**, 1–96.

Mackereth, F.J.H. (1966) Some chemical observations on post-glacial lake sediments. *Phil. Trans. Roy. Soc. Lond.* **250**, 165–213.

McLachlan, A.J. (1970) Some effects of annual fluctuations in water level on the larval chironomid communities in Lake Kariba, *J. Anim. Ecol.* **39**, 70–90.

McLean, R.C. and Ivimey-Cook, W.R. (1956) *Textbook of Theoretical Botany*, Longman, London.

McLean, W.N. (1940) Windermere basin: rainfall, run-off and storage. *Quart. J. R. Met. Soc.* **66**, 337–362.

McLusky, D.S. (1971) *Ecology of Estuaries*, Heinemann, London.

Magnuson, J.J. (1976) Managing with exotics—a game of chance. *Trans. Amer. Fish. Soc.* **105**, 1–9.

Maitland, P.S. (1962) The fauna of the River Endrick in relation to local water use. *Wat. Waste Treatm. J.* **9**, 78–86.

Maitland, P.S. (1966) *The Fauna of the River Endrick*, Blackie, Glasgow.

Maitland, P.S. (1979) *Synoptic Limnology: the Analysis of British Freshwater Ecosystems*, Institute of Terrestrial Ecology, Cambridge.

Maitland, P.S. (1989) *The Genetic Impact of Farmed Atlantic Salmon on Wild Populations*, Nature Conservancy Council, Edinburgh.

Maitland, P.S., East, K. and Morris, K.E. (1983) Ruffe, *Gymnocephalus cernua* (L.), new to Scotland, in Loch Lomond. *Scot. Nat.* **1983**, 7–9.

Maitland, P.S., Lyle, A.A. and Campbell, R.N.B. (1987) *Acidification and Fish in Scottish Lochs*, Institute of Terrestrial Ecology, Grange-over-Sands.

Maitland, P.S. and Penney, M.M. (1967) The ecology of the Simuliidae in a Scottish river. *J. Anim. Ecol.* **36**, 179–206.

Maitland, P.S. and Price, C.E. (1969) *Urocleidus principalis* (Mizelle, 1936), a North American monogenetic trematode new to the British Isles, probably introduced with the largemouth bass *Micropterus salmoides* (Lacepede, 1802). *J. Fish Biol.* **1**, 17–18.

Maitland, P.S. and Smith, I.R. (1987) The River Tay: ecological changes from source to estuary, *Proc. Roy. Soc. Edin.*, **92B**, 373–392.

Mann, K.H. (1958) Occurrence of an exotic oligochaete *Branchiura sowerbyi*, Beddard 1892, in the River Thames. *Nature, Lond.* **182**, 732.

Mann, K.H. (1964) The pattern of energy flow in the fish and invertebrate fauna of the River Thames. *Verh. Int. Ver. Limnol.* **15**, 485–495.

Mann, K.H. (1969) The dynamics of aquatic ecosystems. *Adv. Ecol. Res.* **6**, 1–71.

Mann, K.H., Britton, R.H., Kowalczewski, A., Lack, T.J., Mathews, C.P. and McDonald, I. (1972) Productivity and energy flow at all trophic levels in the River Thames, England. *Proc. I.B.P. Symp., Poland. 1970*, 579–596.

Margalef, R. (1960) Ideas for a synthetic approach to the ecology of running waters. *Int. Rev. Ges. Hydrobiol.* **45**, 133–153.

Marsh, C.M. (1963) Use of water for navigation. *Proc. Symp. Inst. Civil Eng. Lond.* **1963**, 52–58.

Marshall, S.M. and Orr, A.P. (1928) The photosynthesis of diatom cultures in the sea. *J. Mar. Biol. Ass. U.K.* **16**, 321–364.

Mathews, R.O., McAffrey, F. and Hart, E. (1984) Acid rain in Ireland, *Ir. J. Env. Sci.* **1**, 47–50.

Mauna Loa Observatory (1985) *Carbon Dioxide in the Atmosphere 1958–88*, Mauna Loa Observatory, Hawaii.

Meinzer, O.E. (1942) *Hydrology*, Dover Publications, New York.

Mills, D.H. (1972) *An Introduction to Freshwater Ecology*, Oliver & Boyd, Edinburgh.

Milner, A. (1989) Personal communication.

Milner, N.J. Scullion, J., Carling, P.A. and Cusp, D.T. (1981) The effects of discharge on sediment dynamics and consequent effects on invertebrates and salmonids in upland rivers, *Adv. Appl. Biol.*, **6**, 153–220.

Miyadi, D., Kawanabe, H. and Mizuno, N. (1981) *The Freshwater Fishes of Japan*, Hoikusha, Osaka.

Moon, H.P. (1935) Methods and apparatus suitable for an investigation of the littoral region of oligotrophic lakes. *Int. Rev. Hydrobiol.* **32**, 319–333.

Morgan, N.C. and McLusky, D.S. (1974) A summary of the Loch Leven I.B.P. results in relation to lake management and future research. *Proc. Roy. Soc. Edin.* **74**, 407–416.

Morgan, N.C. and Waddell, A.B. (1961) Insect emergence from a small trout loch and its bearing on the food supply of fish. *Freshw. Salm. Fish. Res.* **25**, 1–39.

Mortimer, C.H. (1941) The exchange of dissolved substances between mud and water in lakes. *J. Ecol.* **129**, 280–329.

Mortimer, C.H. (1953) A review of temperature measurement in limnology. *Mitt. Int. Ver. Limnol.* **1953**, 1–25.

Mortimer, C.H. (1959) The oxygen content of air-saturated fresh waters and aids in calculating percentage saturation. *Mitt. Int. Ver. Limnol.* **6**, 1–20.

Muller, K. (1954) Investigations on the organic drift in North Swedish streams. *Rep. Inst. Freshw. Res. Drottningholm* **35**, 133–148.

Muller, K. (1965) Field experiments on periodicity of freshwater invertebrates. *Proc. Feld. Summer School* **1**, 314–317.

Mundie, J.H. (1956) Emergence traps for aquatic insects. *Mitt. Int. Ver. Limnol.* **7**, 1–13.

Muniz, I.P. and Leivestad, H. (1980) Acidification—effects on freshwater fish. In Drablos, D. and Tollan, A. *Ecological Impact of Acid Precipitation*, SNSF project, Oslo, 84–92.

Murray, J. (1910) Characteristics of lakes in general and their distribution over the surface of the globe. *Bath. Surv. Freshw. Lochs Scot.* **1**, 514–618.

Murray, J. and Pullar, L. (1910) *Bathymetrical Survey of the Fresh Water Lochs of Scotland*, Challenger, Edinburgh.

Naumann, E. (1929) The scope and chief problems of regional limnology. *Int. Rev. Ges. Hydrobiol.* **22**, 423–444.

Nielsen, A. (1950) The torrential invertebrate fauna. *Oikos* **2**, 176–196.

Nielsen, E.S. (1952) The use of radioactive carbon (C^{14}) for measuring organic production in the sea. *J. Cons. Int. Explor. Mer* **18**, 117–140.

Nikolsky, G.V. (1963) *The Ecology of Fishes*, Academic Press, London.

Nixon, M. (1963) Flood regulation and river training in England and Wales. *Proc. Symp. Inst. Civil Eng. Lond.* **1963**, 137–150.

Obeng, L.E. (1981) Man's impact on tropical rivers. In: Lock, M.A. and Williams, D.D. (eds) *Perspectives in Running Water Ecology*, Plenum, New York, pp. 265–288.

Odum, H.T. (1956) Primary production in flowing waters, *Limnol. Oceanogr.* **1**, 102–117.

Odum, H.T. (1957) Trophic structure and productivity of Silver Springs, Florida, *Ecol. Monogr.* **27**, 55–112.

Oglesby, R.T., Carlson, C.A. and McCann, J.A. (1972) *River Ecology and Man*, Academic Press, New York.

Ohle, W. (1934) Chemische und Physikalische Untersuchungen Norddeutscher Seen, *Arch. Hydrobiol.* **26**, 386–464.

Overrein, L.N., Seip, H.M. and Tollan, A. (1980) *Acid Precipitation: Effects on Forests and Fish*, SNSF Project, Oslo.

Pardo, G. (1976) Inventario y zonificacion de la cuenca para fines hidroagricolas. *Conf. Foro.* **1976**, D3, 1–7.

Pearsall, W.H. (1917) The aquatic and marsh vegetation of Esthwaite Water. *J. Ecol.* **5**, 189–202.

Pearsall, W.H. (1921) The development of vegetation in the English lakes, considered in relation to the general evolution of glacial lakes and rock basins. *Proc. Roy. Soc. Lond.* **92**, 259–284.

Pearsall, W.H., Gardiner, A.C. and Greenshields, F. (1946) Freshwater biology and water supply in Great Britain, *Sci. Publ. Freshw. Biol. Ass.* **11**, 1–90.

Pennak, R.W. (1955) Comparative limnology of eight Colorado mountain lakes. *Univ. Color. Biol. Ser.* **2**, 1–75.

Pennington, W. (1973) The recent sediments of Windermere. *Freshw. Biol.* **3**, 363–382.

Perl, G. (1935) Zur kenntniss der wahren Sonnenstrahlung in verschieden geographischen Braiten. *Met. Z.* **1935**, 85–89.

Phelps, E.B. (1944) *Stream Sanitation*, Wiley, New York.

Phillipson, J. (1966) *Ecological Energetics*, Arnold, London.

Prickett, C.N. (1963) Use of water for agriculture. *Proc. Symp. Inst. Civil Eng. Lond.* **1963**, 15–29.

Rafferty, S.R. (1963) Introductory survey. *Proc. Symp. Inst. Civil Eng. Lond.* **1963**, 5–8.

Raimondo, P. (1975) Monograph on operation fisheries, *Mopti. CIFA Occ. Pap.* **4**, 294–311.

Rawson, D.S. (1939) Some physical and chemical factors in the metabolism of lakes. *Amer. Assoc. Adv. Sci. Publ.* **10**, 9–26.

Rawson, D.S. (1944) The calculation of oxygen saturation values and their correction for altitude. *Limnol. Soc. Amer. Spec. Publ.* **15**, 1–4.

Rawson, D.S. (1951) The total mineral content of lake waters. *Ecology* **34**, 669–672.

Rawson, D.S. (1955) Morphometry as a dominant factor in the productivity of large lakes, *Verh. Int. Ver. Limnol.* **12**, 164–175.

Reid, G.K. (1961) *Ecology of Inland Waters and Estuaries*, Reinhold, New York.

Reid, G.M. (1987) The Lake Victoria fisheries problem. *Cichlidae* **8**, 1–13.

Reynoldson, T.B. (1958) The quantitative ecology of lake-dwelling triclads in northern Britain. *Oikos* **9**, 94–138.

Reynoldson, T.B., Smith, B.D. and Maitland, P.S. (1981) A species of North American triclad new to Britain found in Loch Ness, Scotland. *J. Zool.* **193**, 531–539.

Ricker, W.E. (1934a) A critical discussion of various measures of oxygen saturation in lakes. *Ecology* **15**, 348–363.

Ricker, W.E. (1934b) An ecological classification of Ontario streams. *Publ. Ont. Fish. Res. Lab.* **49**, 1–114.

Ricker, W.E. (1958) Handbook of computations for biological statistics of fish populations. *Bull Fish. Res. Bd. Can.* **119**, 1–300.

Ricker, W.E. (1968) *Methods for Assessment of Fish Production in Fresh Waters*, Blackwell Scientific Publications, Oxford.

Rigler, F.H. (1956) A tracer study of the phosphorus cycle in lake water. *Ecology* **7**, 550–562.

Rosseland, B.O., Sevaldrud, I., Svalastag, D. and Muniz, I.P. (1980) Studies on freshwater fish populations—effects of acidification on reproduction, population structure, growth and food selection. *Proc. Int. Conf. Ecol. Impact Acid Precip. Sandefjord* **1980**, 336–337.

Rothacher, J.S. (1953) White Hollow Watershed management: 15 years of progress in character of forest, run-off and stream-flow. *J. For.* **51**, 731–738.

Rothschild, L. (1961) *A Classification of Living Animals*, Longman, London.

Round, F.E. (1965) *The Biology of Algae*, Arnold, London.

Rounsefell, G.A. and Everhardt, W.H. (1953) *Fishery Science: Its Methods and Applications*, Wiley, New York.

Royal Commission for Environmental Pollution (1979) *Agriculture and Pollution*, Her Majesty's Stationery Office, London.

Russell-Hunter, W.D. (1968) *A Biology of Lower Invertebrates*, Macmillan, New York.

Russell-Hunter, W.D. (1969) *A Biology of Higher Invertebrates*, Macmillan, New York.

Russell-Hunter, W.D. (1970) *Aquatic Productivity: an Introduction to some Basic Aspects of Oceanography and Limnology*, Macmillan, New York.

Ruttner, F. (1963) *Fundamentals of Limnology*, University Press, Toronto.

Ruttner, F. and Hermann, K. (1937) Uber Temperatur messungen mit einem neue Modell des Lunzer Wasserschopfers. *Arch. Hydrobiol.* **31**, 682–686.

Ryder, R.A. and Kerr, S.R. (1984) Reducing the risk of fish introductions: a rational approach to the management of cold-water communities. *EIFAC Tech. Pap.* **42**, 510–533.

Rzoska, J. (1974) *The Nile: Biology of an Ancient River*, Junk, The Hague.

Saur, J.F.T., Tonolli, V. and Verduin, J. (1964) Sources of limnological and oceanographic apparatus and supplies. *Limnol. Oceanogr.* Suppl. **9**, 1–32.

Savage, W.G. (1966) *Bacteriological Examination of Water Supplies*, London.

Schenck, H. (1886) Vergleichende Anatomie der submersen Gewachse, *Biblthca. Bot.* **1**, 1–67.

Schmitz, W. (1955) Physiographische Aspekte der limnologischen Fliesgewasserntypen. *Arch. Hydrobiol.* **22**, 510–523.

Schnick, I.R.A., Meyer, F.P. and Walsh, D.F. (1986) Status of fishery chemicals in 1985. *Prog. Fish. Cult.* **48**, 1–17.

Schofield, C.L. (1982) Historical fisheries changes in the United States related to decrease in surface water pH. In Johnson R.E., *Acid Rain: Fisheries*, Cornell University, New York, 57–61.

Sculthorpe, C.D. (1967) *The Biology of Aquatic Vascular Plants*, Arnold, London.

Sioli, H. (1975) Amazon tributaries and drainage basins. *Ecol. Stud.* **10**, 199–213.

Slack, H.D. (1954) The bottom deposits of Loch Lomond. *Proc. Roy. Soc. Edin.* **65**, 213–238.

Slack, H.D. (1955) A quantitative plankton net for horizontal sampling. *Hydrobiologia* **7**, 264–268.

Slack, H.D. (1957) Physical and chemical data, *Glasg. Univ. Publ., Stud. Loch Lomond*, **1**, 14–26.

Slobodkin, L.B. (1962) Energy in animal ecology. *Adv. Ecol. Res.* **1**, 69–101.

Smith, G.M. (1951) *Manual of Phycology*, Chronica Botanica, Waltham.

Smith, I.R. (1974) The structure and physical environment of Loch Leven, Scotland. *Proc. Roy. Soc. Edin.* **74**, 81–100.

Smith, I.R., Lyle, A.A. and Rosie, A.J. (1981) Comparitive physical limnology, *Monogr. Biol.* **44**, 29–65.

Smith, B.D., Maitland, P.S. and Ptnnock S.M. (1987) A comparative study of water level regimes and littoral benthic communities is Scottish lochs. *Biol. Conserv.* **39**, 291–316.

Smith, S.H. (1968) Species succession and fishery exploitation in the Great Lakes. *J. Fish. Res. Bd. Can.* **25**, 667–693.

Sorokin, Y.I. (1964) *Handbook on Microbial Production*, Blackwell Scientific Publications, Oxford.

Sorokin, V.I. and Kadota, H. (1972) *Handbook on Microbial Production*, Blackwell Scientific Publications, Oxford.

Spence, D.H.N. (1967) Factors controlling the distribution of freshwater macrophytes with particular reference to the lochs of Scotland. *J. Ecol.* **55**, 147–170.

Spence, D.H.N., Campbell, R.M. and Chrystal, J. (1971) Spectral intensity in some Scottish freshwater lochs. *Freshw. Biol.* **1**, 321–337.

Stenmark, A. and Malmberg, O. (1987) *Parasites and Diseases in Natural Waters and Aquaculture in Nordic Countries*, University of Stockholm, Stockholm.

Stewart, W.J.P., Fitzgerald, G.P. and Burns, R.H. (1967) In situ studies on N_2 fixation using the acetylene reduction technique, *Proc. Nat. Acad. Sci. U.S.* **58**, 2071–2078.

Stott, B. (1977) On the question of the introduction of the grass carp (*Ctenopharyngodon idella* Val.) into the United Kingdom. *Fish. Mgt.* **3**, 63–71.

Strom, K.M. (1945) The temperature of maximum density in fresh waters. *Geofys. Puld.* **16**, 1–14.

Stuart, T.A. (1953) Water currents through permeable gravels and their significance to spawning salmonids, etc. *Nature, Lond.* **172**, 407–408.

Stuart, T.A. (1959) The influence of drainage works, levees, dykes, dredging, etc. on the aquatic environment and stocks. *Proc. IUCN Tech. Meet. Athens* **4**, 337–345.

Stumm, W. and Morgan, J. (1981) *Aquatic Chemistry*, Wiley, New York.

Sutcliffe, D.W. and Carrick, T.R. (1973) Studies on mountain streams in the English Lake District. 1. pH, calcium and the distribution of invertebrates in the River Duddon. *Freshw. Biol.* **3**, 437–462.

Talling, J.F. (1957) The phytoplankton population as a compound photosynthetic system. *New Phytol.* **56**, 133–149.

Talling, J.F. (1971) The underwater light climate as a controlling factor in the production ecology of freshwater phytoplankton. *Mitt. Int. Ver. Limnol.* **19**, 214–243.

Tansley, A.G. (1939) *The British Islands and their Vegetation*, University Press, Cambridge.

Taylor, E.W. (1958) *The Examination of Water and Water Supplies*, Thresh, Beale & Suckling, London.

Teal, J.M. (1957) Community metabolism in a temperate cold spring. *Ecol. Monogr.* **27**, 283–302.

Teal, J.M. (1962) Energy flow in the salt marsh ecosystem of Georgia. *Ecology* **43**, 614–624.

Thienemann, A. (1912) Der. Bergbach des Sauerlands. Faunistisch-biologische Untersuchungen. *Hydrobiol.* Suppl. **4**, 1–125.

Thienemann, A. (1925) *Die Binnengewasser Mitteleuropas*. Stuttgart.

Tiffany, L.H. (1938) *Algae*, Springfield, Illinois.

Tiffany, L.H. (1951) Ecology of Freshwater Algae. In Smith, G.M., *Manual of Phycology*, Chronica Botanica, Waltham.

Tinbergen, N. (1953) *Social Behaviour in Animals*, Methuen, London.

Toha, J.C. and Jaques, I. (1989) Algal solution? *New Sci.* **1691**, 71.

Tome, M.A. and Pough, F.H. (1982) Responses of amphibians to acid precipitation. In Johnson, R.E., *Acid Rain: Fisheries*, Cornell University, New York.

Tonolli, V. (1949) Ripartizione Spaziale e Migraziore verticale dello Zooplancton—Ricerche e Considerazione. *Mem. Ist. Ital. Idrobiol.* **5**, 211–228.

Tressler, W.L., Tiffany, L.H. and Spencer, W.P. (1940) Limnological studies of Buckeye Lake, Ohio. *Ohio. J. Sci.* **40**, 261–290.

United Kingdom Acid Waters Review Group (1986) *Acidity in United Kingdom Fresh Waters*. Department of the Environment, London.

Van Dorn, W.G. (1956) Large volume water samplers. *Trans. Amer. Geophys. Un.* **37**, 682–684.

Vollenweider, R.A. (1971) *A Manual on Methods for Measuring Primary Production in Aquatic Environments*, Blackwell Scientific Publications, Oxford.

Wedderburn, E.M. (1910) Temperature of Scottish lakes. *Bath. Surv. Freshw. Lochs Scot.* **1**, 91–144.

Wedderburn, E.M. (1912) Temperature observations in Loch Earn, with a further contribution to the hydrodynamical theory of the temperature seiche. *Trans. Roy. Soc. Edin.* **48**, 629–695.

Welch, P.S. (1948) *Limnological Methods*, Blakiston, New York.

Welch, P.S. (1952) *Limnology*, McGraw-Hill, New York.

Welcomme, R.L. (1979) *Fishery Ecology of Floodplain Rivers*, Longman, London.

Wesenberg-Lund, C. (1971) Furesostudier. *K. Danske Vidensk. Selsk.* **8**.

West, G.S. (1916) *Algae*, Cambridge.

Westlake, D.F. (1966) The light climate for plants in rivers. *Brit. Ecol. Symp.* **6**, 99–119.

Whipple, G.C. (1898) Classification of lakes according to temperature. *Amer. Nat.* **32**, 25–53.

Whipple, G.C. (1927) *The Microscopy of Drinking Water*, Wiley, New York.

Whittaker, R.H. (1970) *Communities and Ecosystems*, Macmillan, New York.

Williams, W.D. (1981) Running water ecology in Australia. In Lock, M.A. and Williams, D.D., *Perspectives in Running Water. Ecology*, Plenum, New York, 367–392.

Willoughby, L.G. (1971) Observations on some aquatic Actinomycetes of streams and rivers. *Freshw. Biol.* **1**, 23–27.

Wood, R.B., Prosser, M.V. and Baker, R.M. (1969) A seasonal study of stratification in a tropical African lake at 1800 m altitude. *Verh. Int. Ver. Limnol.* **17**, 1050–1051.

Wright, R.F. and Snekvic, E. (1978) Acid precipitation—chemistry and fish populations in 700 lakes in southernmost Norway. *Verh. Int. Ver. Limnol.* **20**, 765–775.

Young, J.Z. (1955) *The Life of Vertebrates*, University Press, Oxford.

Young, J.Z. (1960) *The Life of Mammals*, University Press, Oxford.

Index